SRA Number Worlds

Number Patterns and Relationships

Unit 2 Workbook
Level G

Author
Sharon Griffin
Associate Professor of Education and
Adjunct Associate Professor of Psychology
Clark University
Worcester, Massachusetts

Building Blocks Authors

Douglas H. Clements
Professor of Early Childhood
and Mathematics Education
University at Buffalo
State University of New York, New York

Julie Sarama
Associate Professor of Mathematics Education
University at Buffalo
State University of New York, New York

Contributing Writers
Sherry Booth, Math Curriculum Developer, Raleigh, North Carolina
Elizabeth Jimenez, English Language Learner Consultant, Pomona, California

Program Reviewers

Jean Delwiche
Almaden Country School
San Jose, California

Cheryl Glorioso
Santa Ana Unified School District
Santa Ana, California

Sharon LaPoint
School District of Indian River County
Vero Beach, Florida

Leigh Lidrbauch
Pasadena Independent School District
Pasadena, Texas

Dave Maresh
Morongo Unified School District
Yucca Valley, California

Mary Mayberry
Mon Valley Education Consortium, AIU 3
Clairton, Pennsylvania

Lauren Parente
Mountain Lakes School District
Mountain Lakes, New Jersey

Juan Regalado
Houston Independent School District
Houston, Texas

M. Kate Thiry
Dublin City School District
Dublin, Ohio

Susan C. Vohrer
Baltimore County Public Schools
Baltimore, Maryland

SRAonline.com

McGraw Hill SRA

Copyright © 2007 SRA/McGraw-Hill.

All rights reserved. Except as permitted under the United States Copyright Act, no part of this publication may be reproduced or distributed in any form or by any means, or stored in a database or retrieval system, without the prior written permission of the publisher, unless otherwise indicated.

Printed in the United States of America.

Send all inquiries to:
SRA/McGraw-Hill
8787 Orion Place
Columbus, OH 43240-4027

R53245.01

1 2 3 4 5 6 7 8 9 QPD 12 11 10 09 08 07 06

Photo Credits
2-39 ©PhotoDisc/Getty Images, Inc.

The McGraw-Hill Companies

Practice
Replace the question mark with the correct symbol to make a true statement.

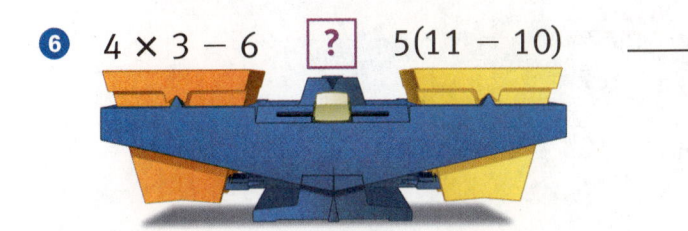

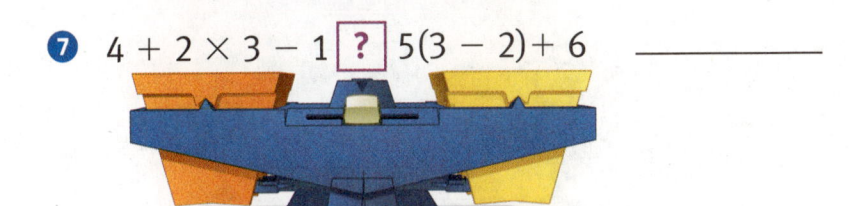

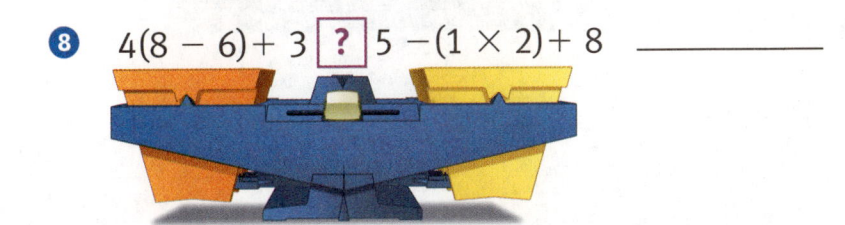

4) $2 + 1 \times 7$? $3(4 - 2)$ _____

5) $2(10 - 5)$? $6 + 4$ _____

6) $4 \times 3 - 6$? $5(11 - 10)$ _____

7) $4 + 2 \times 3 - 1$? $5(3 - 2) + 6$ _____

8) $4(8 - 6) + 3$? $5 - (1 \times 2) + 8$ _____

Reflect
Write two expressions that would balance a scale when you put one on each side.

_____

Equality • Lesson 3

Practice
Create a graph for the bus-pass function.

Input (number of bus rides)	Output (bus-pass balance)
0	$20.00
1	$19.50
2	$19.00
3	$18.50
4	$18.00
5	$17.50

Reflect
Do you think it is possible to determine a function rule just by looking at the graph? Explain and give an example.

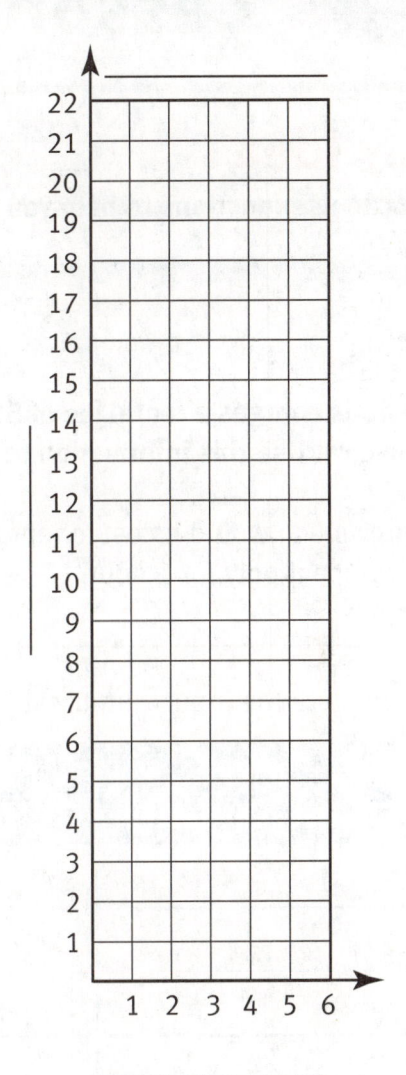

_____
_____
_____
_____

Functional Relationships • Lesson 2

Week 4

More with Functional Relationships

Lesson 4

Key Idea
Relationships exist between equations and graphs.

Try This
Complete the input/output table for the equation below.

△ = ◯ + 3

Input ◯	Output △
1	
2	
3	
4	
5	
6	

❶ Plot the values from the table on the coordinate grid.

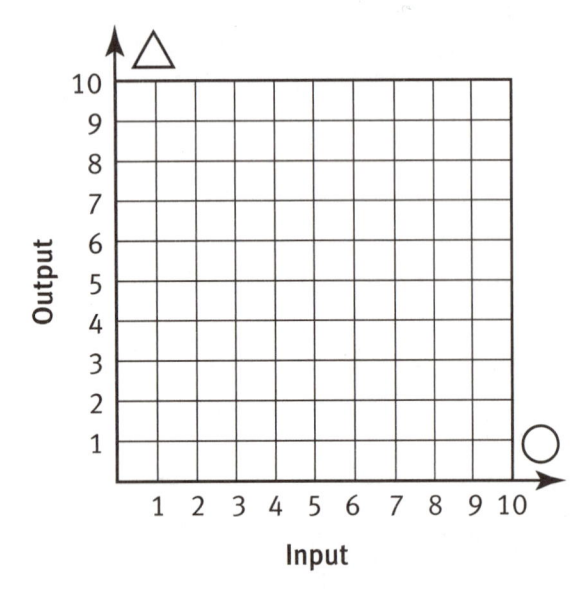

△
10
9
8
7
6
5
4
3
2
1

Output

1 2 3 4 5 6 7 8 9 10 ◯
Input

38   Number Patterns and Relationships • Week 4

Practice
Find each unknown cost of the items purchased from a hardware store.

4 🔨 + 📦 📦 = $17

Each box of nails costs $2.50. What is the cost of the hammer?

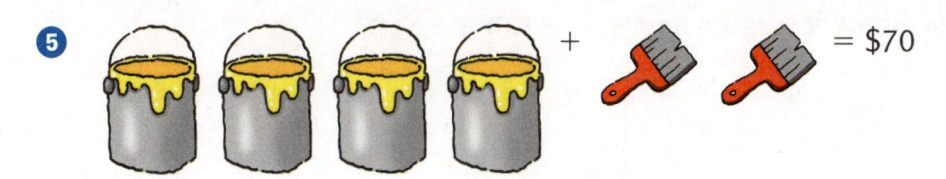

_____

5 🪣🪣🪣🪣 + 🖌🖌 = $70

Each can of paint costs $15. What is the cost of each paintbrush?

_____

6 🔩🔩 + 📏📏 = $80

Each drill costs $32. What is the cost of each tape measure?

_____

Reflect
Suppose Carl spent $1.05 at the bookstore for two folders and one marker. If the cost of a folder is the same as the cost of a marker, what is the cost of each item? Explain.

_____

_____

Variables • Lesson 2

Practice

The shapes represent two different variables. Complete the table for each equation.

○ + △ = 15

○	△
3	
5	
	9
	7
10	
	4

Problems 7–12

□ + □ + ◹ = 20

□	◹
2	
3	
	10
	8
7	
	2

Problems 13–18

Reflect

What is another set of values for the square and triangle that would make a true number sentence?

_____

Variables • Lesson 3

Practice

Write an equation for each area model where the length and width are known. Then write an equation using a variable to represent the hidden columns.

4

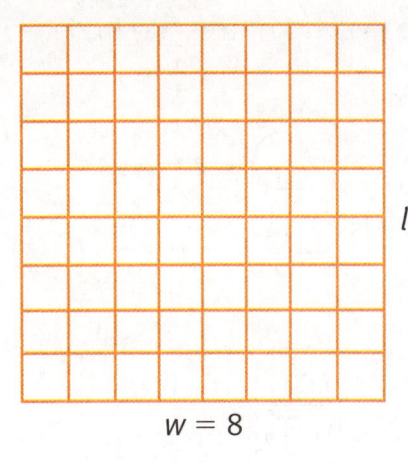

$l = 8$
$w = 8$

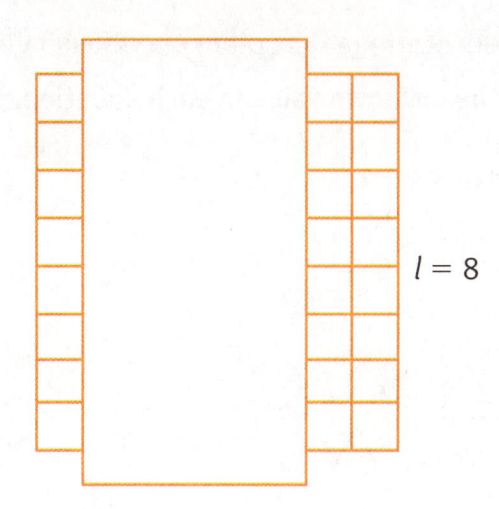

$l = 8$

5

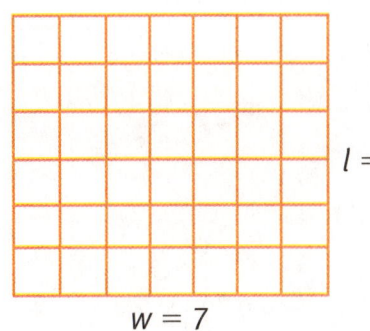

$l = 6$
$w = 7$

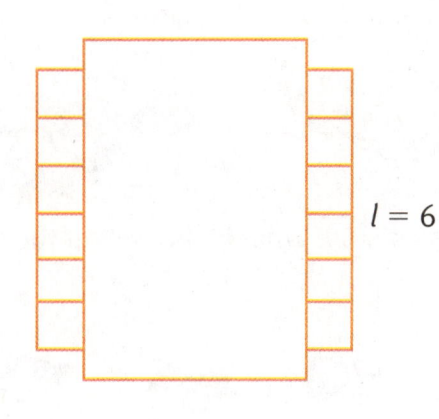

$l = 6$

Reflect

How could you write an area equation from a model if a certain number of rows is unknown? Give an example of such an equation.

Variables • Lesson 4

Practice
Replace the question mark with the correct symbol to make a true statement.

4   2 + 1 × 7   **?**   3(4 − 2)   _____

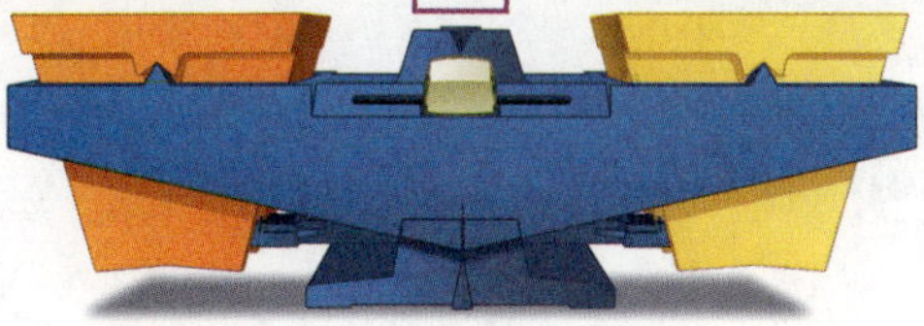

5   2(10 − 5)   **?**   6 + 4   _____

6   4 × 3 − 6   **?**   5(11 − 10)   _____

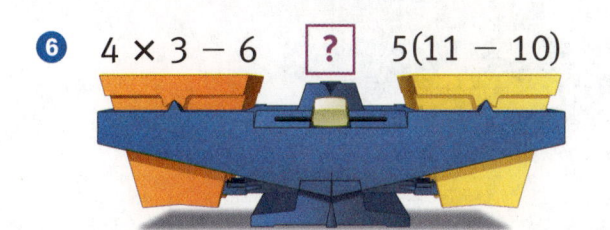

7   4 + 2 × 3 − 1   **?**   5(3 − 2) + 6   _____

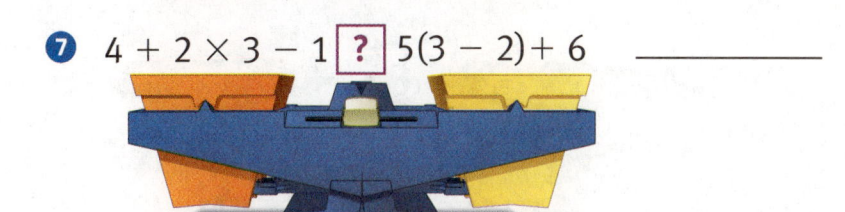

8   4(8 − 6) + 3   **?**   5 − (1 × 2) + 8   _____

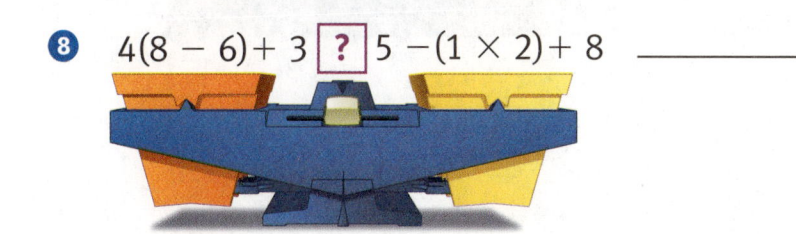

Reflect
Write two expressions that would balance a scale when you put one on each side.

_____

Equality • Lesson 3

Week 1 — Variables

Lesson 1

Key Idea

A variable is a letter or symbol, such as n, x, or □, that stands for a value in an expression or an equation.

An equation is a number sentence stating that two expressions are equal.

$$8 + 3 = 11 \qquad 10 - 5 = 5 \qquad 6 - 1 = 2 + 3$$

When an equation contains variables, you can show these unknown values with symbols or with letters.

$$\square + \square = 8 \qquad x - 4 = 5 \qquad \triangle + 1 = 12$$

Try This

Find the unknown value in each equation. Substitute values into the equation until you have a true number sentence.

1. $\square + 3 = 8 - 1$
What is $\square$?
_____

2. $\triangle + 1 = 7 + 8$
What is $\triangle$?
_____

3. $14 - \triangle = 5 + 3$
What is $\triangle$?
_____

4. $12 - \bigcirc = 10$
What is $\bigcirc$?
_____

5. $\bigcirc + 4 = 10 - 2$
What is $\bigcirc$?
_____

6. $11 - 6 = \square + 2$
What is $\square$?
_____

Week 1 Variables

Lesson 2

Key Idea
Variables can represent money amounts.

Try This

Ms. O'Brien works at the school bookstore. She sold the following items to students during lunch. Find the unknown costs of the items.

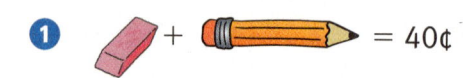

1 eraser + pencil = 40¢

The eraser costs 25¢. What is the cost of the pencil?

_____

2 marker + marker + folder = 95¢

Each marker costs 30¢. What is the cost of the folder?

_____

3 paper clip + paper clip + paper clip + paper clip + paper clip + paper clip + pen = 92¢

Each paper clip costs 2¢. What is the cost of the pen?

_____

4 Number Patterns and Relationships • Week 1

Practice

The shapes represent two different variables. Complete the table for each equation.

$$\bigcirc + \triangle = 15$$

○	△
7. 3	12
8. 5	10
9. 6	9
10. 8	7
11. 10	5
12. 11	4

$$\square + \square + \triangle = 20$$

□	△
13. 2	16
14. 3	14
15. 5	10
16. 6	8
17. 7	6
18. 9	2

Reflect

What is another set of values for the square and triangle that would make a true number sentence?

_____

Variables • Lesson 3 7

Week 1 — **Variables**

Lesson 4

> **Key Idea**
> In this lesson, you will continue to explore equations by using area models.

Try This

The model on the left shows a rectangle with an area of 63 square units. Use the model to answer each question.

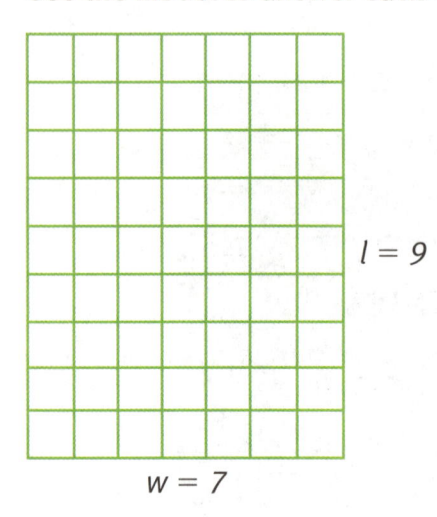

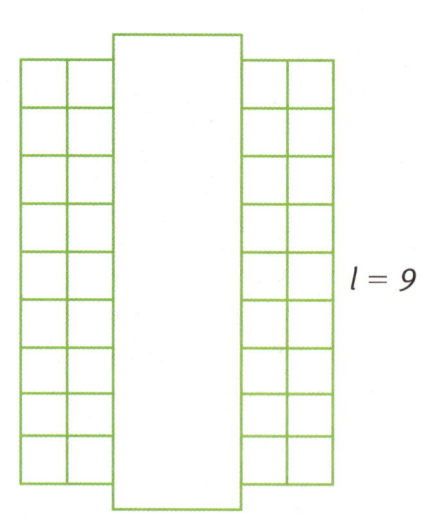

l = 9
w = 7

l = 9

1 How many columns of squares are not hidden in the second figure?

_____

2 Let △ represent the number of columns that are hidden in the second figure. Write an expression for the length.

_____

3 Write an equation for the area of the second figure with the variable △.

_____

8 Number Patterns and Relationships • Week 1

Practice

Write an equation for each area model where the length and width are known. Then write an equation using a variable to represent the hidden columns.

4

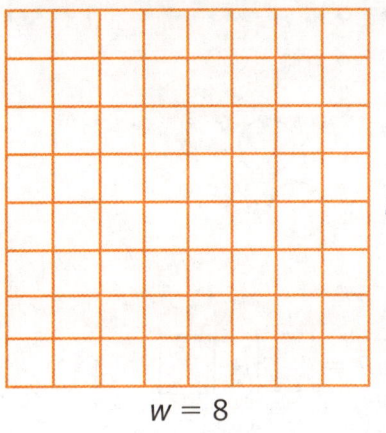

$l = 8$
$w = 8$

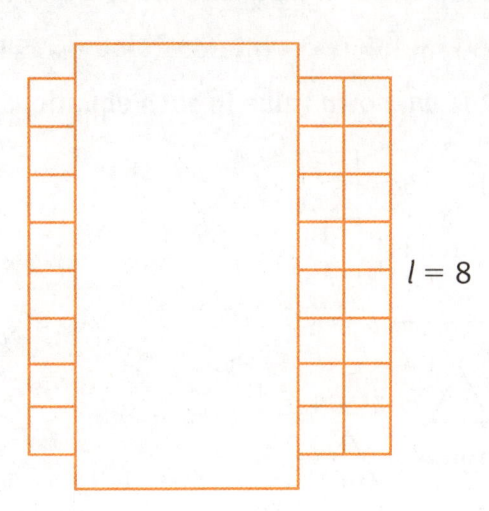

$l = 8$

5

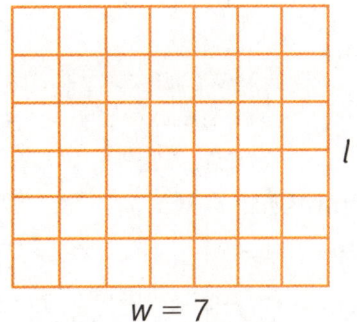

$l = 6$
$w = 7$

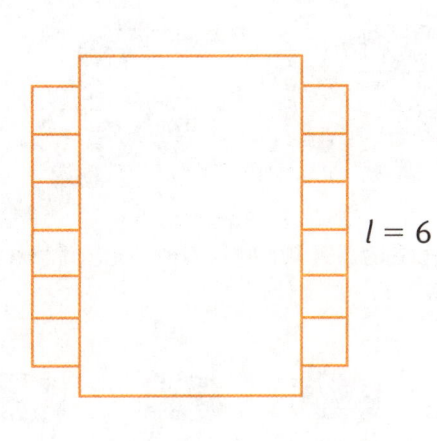

$l = 6$

Reflect

How could you write an area equation from a model if a certain number of rows is unknown? Give an example of such an equation.

Variables • Lesson 4   9

4

_____

5

_____

Reflect
In Exercise 5, explain how you determined your answer.

_____
_____
_____
_____

Equality • Lesson 2   15

8 $\square - 3 \times 2$ $4 + \bigcirc$

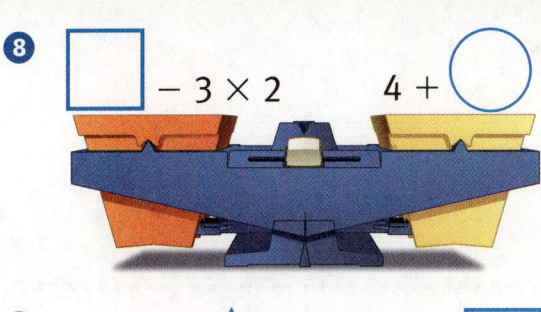

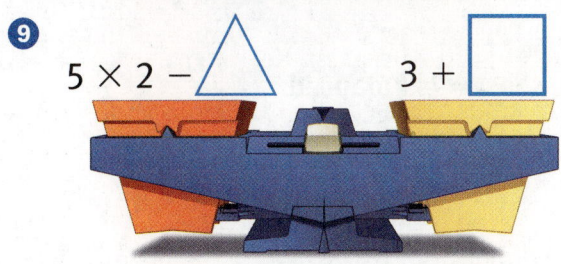

9 $5 \times 2 - \triangle$ $3 + \square$

Fill in the table with values that will set the scale so the right side is greater than the left side.

$\bigcirc + 3 \times 2 \; > \; 2 + \square$

	$\bigcirc$	$\square$
10	1	
11	3	
12		8
13		9
14	7	
15		13

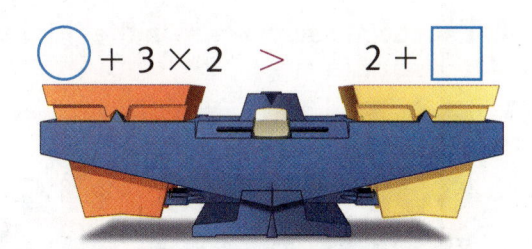

Reflect
Refer to the balance scale used in Problems 10–15. What numbers could you replace each variable with so the scale would represent a "<" inequality?

Equality • Lesson 4   19

Week 3

Functional Relationships

Lesson 5 Review

This week you explored functions and function patterns. You studied input/output tables and graphed the data from input/output tables.

Lesson 1 Tickets to the state fair cost $15 each.

1 Complete the input/output table.

Input (number of tickets)	Output (total cost)
1	
2	
3	
4	
5	

Lesson 2

2 Graph the data from the input/output table above.

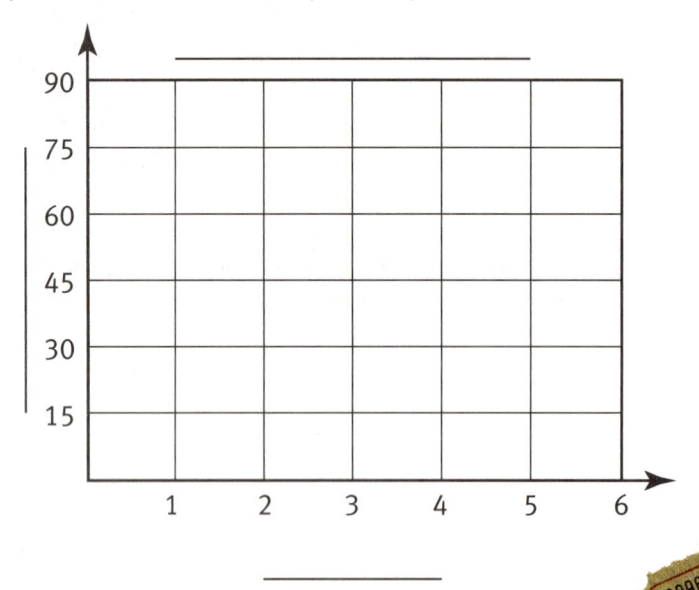

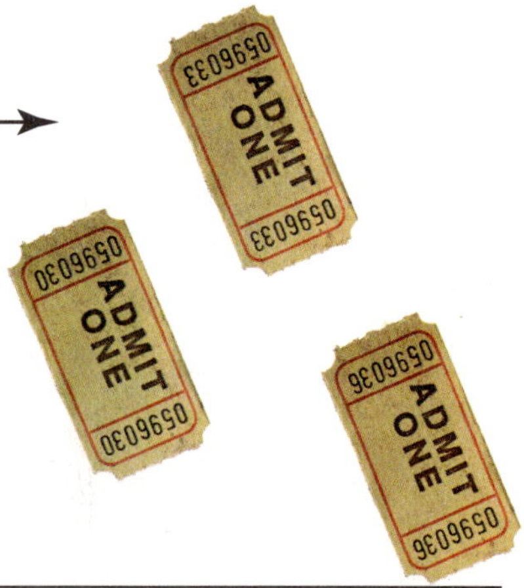

Reflect

What function rule can be used to find the total cost of t tickets from the function above?

30 Number Patterns and Relationships • Week 3

Week 4

More with Functional Relationships

Lesson 5 Review

This week you learned more about functional relationships. You discovered that geometric patterns can be represented as functions. You learned that relationships exist between equations and graphs.

Lesson 1   Use the table below to answer each question.

Car Rental	
Company A	**Company B**
• Up-front fee: $25	• Up-front fee: $0
• Rental fee: $15/day	• Rental fee: $20/day

1. Write a function rule for the total cost of renting a car from Company A for d days.

   _____

2. Write a function rule for the total cost of renting a car from Company B for d days.

   _____

3. Suppose Miss Wilson needs to rent a car for 7 days. Which company will be less expensive? Explain.

   _____
   _____

Lesson 2   Use the pattern of blocks below to answer each question.

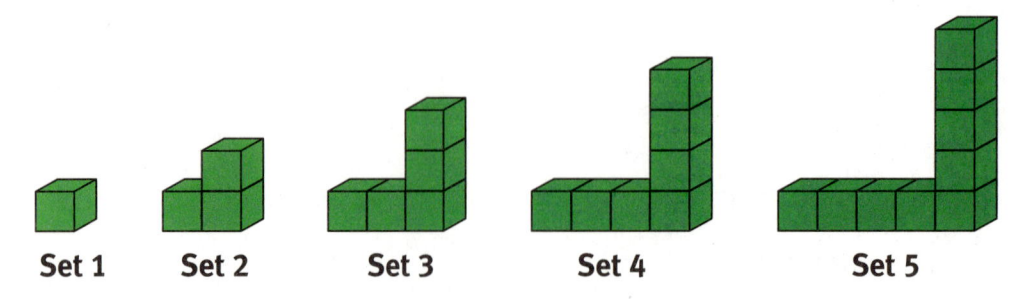

Set 1   Set 2   Set 3   Set 4   Set 5

4. How many blocks were used to create each of the five figures?

   _____

40   Number Patterns and Relationships • Week 4

Week 3 Functional Relationships

Practice

1 Movie tickets to the cinema complex cost $8 each. Complete the input/output table.

Input (number of tickets)	Output (total cost)
1	
2	
3	
4	
5	

2 Graph the data from the input/output table.

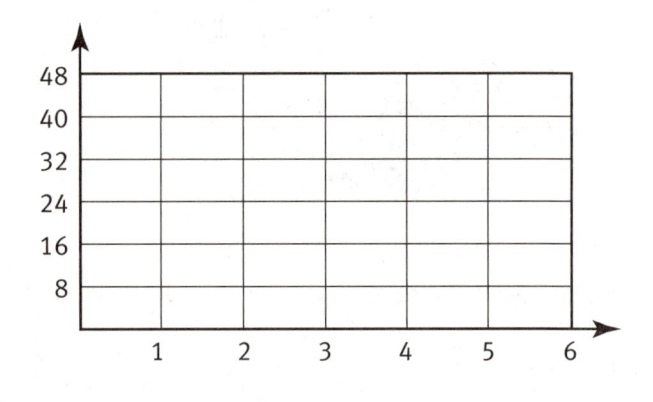

Use the graph to answer the questions below.

3 How would you describe the pattern shown in the graph?

_____

4 What function rule can you use to find the cost of *t* tickets?

_____

5 How much was spent if you have 12 tickets?

_____

6 Is the graph showing a shrinking or growing pattern?

_____

7 If each ticket costs $9, what function rule can you use to find the amount spent on *t* tickets?

_____

44 Number Patterns and Relationships • Week 3 Practice

Week 1 — Variables

Lesson 5 Review

This week you used variables in equations. You chose values that could replace a variable and make a true number sentence.

Lesson 1 Find the unknown value in each equation.

1. $10 - 5 = \square + 2$
   What is $\square$?

2. $11 + 1 = \bigcirc + 2$
   What is $\bigcirc$?

3. $\triangle - 7 = 4 + 1$
   What is $\triangle$?

4. $7 + 9 = \square + \square$
   What is $\square$?

Lesson 2

5. pan + two oven mitts = $15
   The pan costs $9. What is the cost of the two oven mitts?

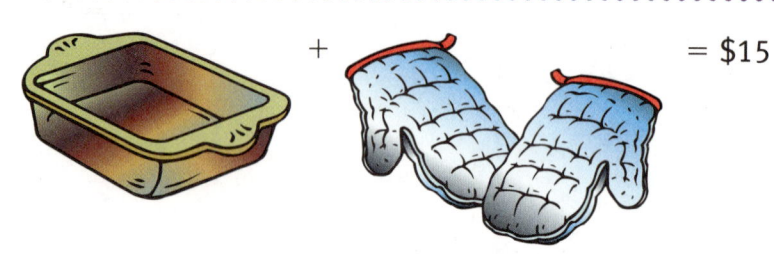

6. 4 chairs + table = $250
   Each chair costs $25. What is the cost of the table?

Reflect

To go along with the table and chairs, Susan purchased a fifth chair and a vase of flowers for the table. Her total purchase cost was $295. Write an equation to represent her purchase using v for the price of the vase. How much did the vase of flowers cost?

Number Patterns and Relationships • Week 1

Lesson 3 Complete the table for each equation below.

$$\bigcirc + \triangle = 8 - 2$$

○	△
1	
2	
	3
	0

7. ○ = 1
8. ○ = 2
9. △ = 3
10. △ = 0

Lesson 4 Write an equation for each area model where the length and width are known. Then write an equation using a variable to represent the hidden rows.

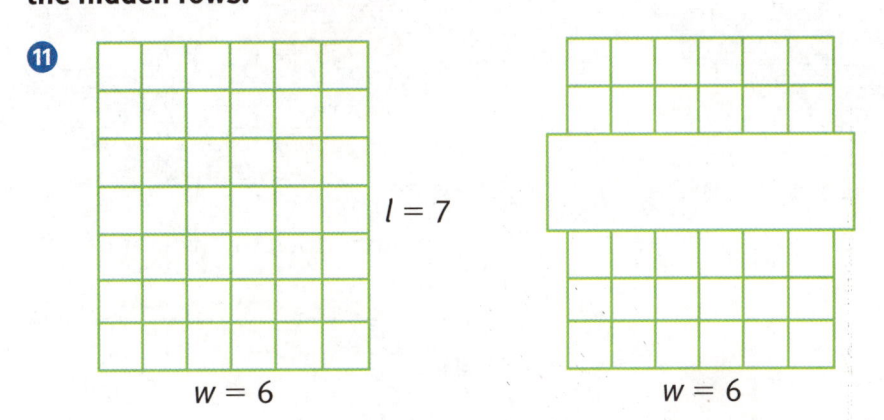

11.

$l = 7$
$w = 6$

$w = 6$

Reflect

In Problem 11 how did you decide on which expression to use for the width of the rectangle that has hidden rows? How many rows are hidden? Show your work.

Variables • Lesson 5 Review **11**

Practice
Tell how much each item might weigh for the seesaw to be balanced.

5

6

7

8

Reflect
Suppose Maria is holding her dog on one side of a seesaw. On the other side is her older brother Kyle. How much might Maria, the dog, and Kyle weigh for the seesaw to be balanced? Explain.

Equality • Lesson 1   13

④

⑤

Reflect
In Exercise 5, explain how you determined your answer.

Equality • Lesson 2

Practice
Replace the question mark with the correct symbol to make a true statement.

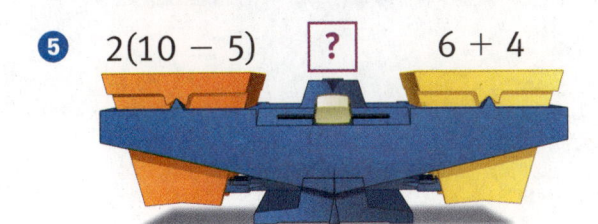

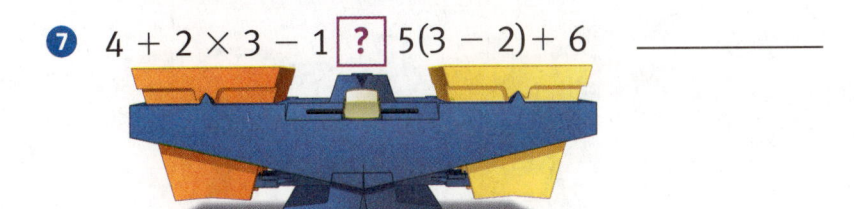

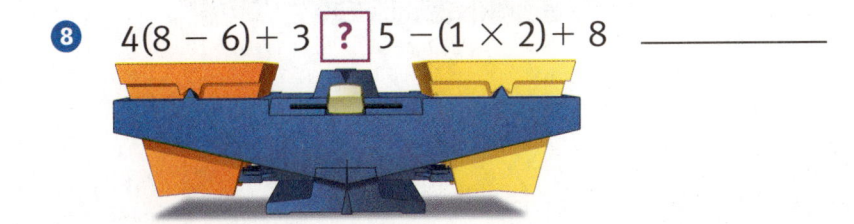

4 $2 + 1 \times 7$? $3(4 - 2)$ _____

5 $2(10 - 5)$? $6 + 4$ _____

6 $4 \times 3 - 6$? $5(11 - 10)$ _____

7 $4 + 2 \times 3 - 1$? $5(3 - 2) + 6$ _____

8 $4(8 - 6) + 3$? $5 - (1 \times 2) + 8$ _____

Reflect
Write two expressions that would balance a scale when you put one on each side.

_____

Equality • Lesson 3   17

Week 2 Equality

Lesson 5 Review

This week you explored balance in equations. You used balanced scales and seesaws to determine the value of a known object or expression.

Lesson 1

Practice
Tell how much each item might weigh so the seesaw is balanced.

1

Lesson 2 What shape(s) can you put on the third scale to balance it?

2

Reflect
Draw three balance scales like those in Problem 2. Use three different types of fruit. Write your answer, and show your work.

20 Number Patterns and Relationships • Week 2

Lesson 3 Replace the question mark with the correct symbol to make a true statement.

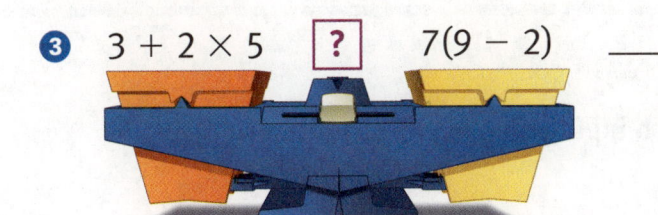

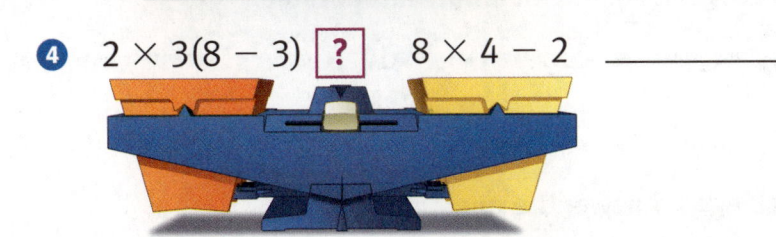

③ 3 + 2 × 5   ?   7(9 − 2)   _____

④ 2 × 3(8 − 3)   ?   8 × 4 − 2   _____

Lesson 4 Fill in the table with values that will balance the scale.

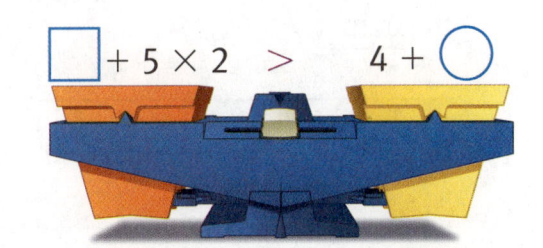

☐ + 5 × 2   >   4 + ◯

☐	◯
1	
4	
	11
	12
8	
	16

Reflect

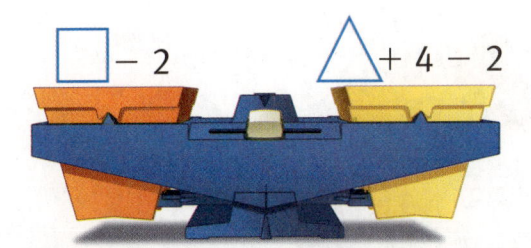

☐ − 2   △ + 4 − 2

☐	△

Create a table like the one in Problems 5–10 for the balance scale above.

Equality • Lesson 5 Review

Week 4

More with Functional Relationships

Lesson 1

Key Idea

You can compare two functions and use them to help you make a decision.

Try This

Students at Richfield High School are planning a dance for their school fund-raiser. Their goal is to raise $1,000. They need to decide whether they want to hire a DJ for the dance or have a live band.

- The DJ is a student at the school and will spin for $75. If a DJ is hired, admission to the dance will be $7.50 per student.

- It will cost $250 to hire a live band. If a live band is hired, admission to the dance will be $10 per student.

❶ Suppose the DJ is hired. After the DJ is paid, how much money will be raised if 10 tickets are sold? 20 tickets? 30 tickets?

_____

❷ Suppose the live band is hired. After the band is paid, how much money will be raised if 10 tickets are sold? 20 tickets? 30 tickets?

_____

❸ Complete the input/output table for each option.

DJ Option	
Input (tickets sold)	Output (money raised)
10	$0
20	
30	
40	
50	
60	

Live Band Option	
Input (tickets sold)	Output (money raised)
10	
20	
30	
40	
50	
60	

32   Number Patterns and Relationships • Week 4

Week 4

More with Functional Relationships

Lesson 2

> **Key Idea**
> You can represent a geometric pattern as a function.

Try This

Grandma Rawls makes patchwork quilts. Each quilt has a row of yellow patches around the perimeter. In the center of the quilts are square red and white patches. The first four sizes of quilts are shown below.

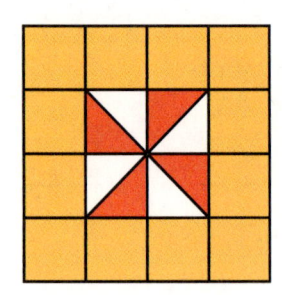

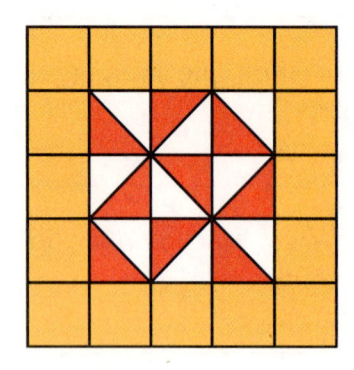

Size 1   Size 2   Size 3   Size 4

1 How many square patches are used in each of the first three sizes?

_____

2 How many yellow patches are used in the borders of each size?

_____

34   Number Patterns and Relationships • Week 4

Practice
Use the quilt designs to answer each question.

3 Complete the input/output table for the number of yellow squares used to create the border of each quilt.

Input (size)	Output (yellow border patches)
1	8
2	
3	
4	
5	
6	

4 What pattern do you notice in the table?

_____

_____

5 What function rule can be used to find the number of yellow border patches needed for a size n quilt?

_____

6 How many yellow border patches would Grandma Rawls need for a size-10 quilt?

_____

Reflect
What size quilt would have 36 yellow border patches? Show your work.

_____

_____

_____

More with Functional Relationships • Lesson 2

Week 4

More with Functional Relationships

Lesson 3

Key Idea
You can represent a geometric pattern as a function.

Try This
Use the quilt patterns again to answer each question.

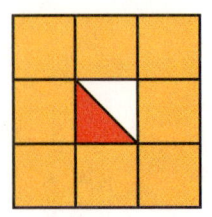

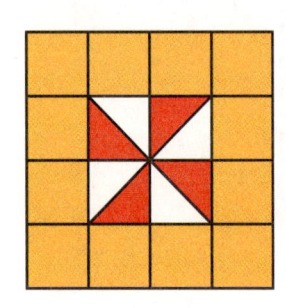

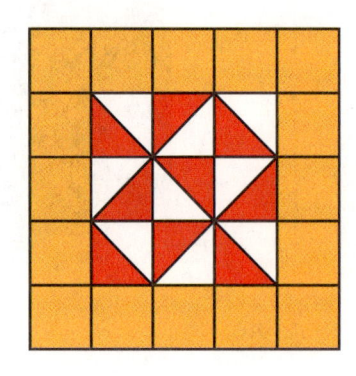

Size 1   Size 2   Size 3   Size 4

1. How many patches are used in each of the first four sizes?
   _____

2. How many red/white square patches are used in each size?
   _____

36   Number Patterns and Relationships • Week 4

Unit 2 Workbook
Level G

SRA
NUMBER WORLDS

Number Patterns and Relationships

featuring
Building Blocks
Software

Author

Sharon Griffin
Associate Professor of Education and
Adjunct Associate Professor of Psychology
Clark University
Worcester, Massachusetts

Building Blocks Authors

Douglas H. Clements
Professor of Early Childhood
and Mathematics Education
University at Buffalo
State University of New York, New York

Julie Sarama
Associate Professor of Mathematics Education
University at Buffalo
State University of New York, New York

Contributing Writers

Sherry Booth, Math Curriculum Developer, Raleigh, North Carolina
Elizabeth Jimenez, English Language Learner Consultant, Pomona, California

Program Reviewers

Jean Delwiche
Almaden Country School
San Jose, California

Cheryl Glorioso
Santa Ana Unified School District
Santa Ana, California

Sharon LaPoint
School District of Indian River County
Vero Beach, Florida

Leigh Lidrbauch
Pasadena Independent School District
Pasadena, Texas

Dave Maresh
Morongo Unified School District
Yucca Valley, California

Mary Mayberry
Mon Valley Education Consortium, AIU 3
Clairton, Pennsylvania

Lauren Parente
Mountain Lakes School District
Mountain Lakes, New Jersey

Juan Regalado
Houston Independent School District
Houston, Texas

M. Kate Thiry
Dublin City School District
Dublin, Ohio

Susan C. Vohrer
Baltimore County Public Schools
Baltimore, Maryland

SRAonline.com

McGraw Hill SRA

Copyright © 2007 SRA/McGraw-Hill.

All rights reserved. Except as permitted under the United States Copyright Act, no part of this publication may be reproduced or distributed in any form or by any means, or stored in a database or retrieval system, without the prior written permission of the publisher, unless otherwise indicated.

Printed in the United States of America.

Send all inquiries to:
SRA/McGraw-Hill
8787 Orion Place
Columbus, OH 43240-4027

R53245.01

1 2 3 4 5 6 7 8 9   QPD   12 11 10 09 08 07 06

Photo Credits

2–39 ©PhotoDisc/Getty Images, Inc.

Contents

Number Patterns and Relationships

Week 1 Variables 2

Week 2 Equality 12

Week 3 Functional Relationships 22

Week 4 More with Functional Relationships 32

Week 1 Practice 42

Week 2 Practice 43

Week 3 Practice 44

Week 4 Practice 45

Week 1 — Variables

Lesson 1

Key Idea

A variable is a letter or symbol, such as n, x, or □, that stands for a value in an expression or an equation.

An equation is a number sentence stating that two expressions are equal.

$$8 + 3 = 11 \qquad 10 - 5 = 5 \qquad 6 - 1 = 2 + 3$$

When an equation contains variables, you can show these unknown values with symbols or with letters.

$$□ + □ = 8 \qquad x - 4 = 5 \qquad △ + 1 = 12$$

Try This

Find the unknown value in each equation. Substitute values into the equation until you have a true number sentence.

1. □ + 3 = 8 − 1

What is □ ?

_____

2. △ + 1 = 7 + 8

What is △ ?

_____

3. 14 − △ = 5 + 3

What is △ ?

_____

4. 12 − ○ = 10

What is ○ ?

_____

5. ○ + 4 = 10 − 2

What is ○ ?

_____

6. 11 − 6 = □ + 2

What is □ ?

_____

2 Number Patterns and Relationships • Week 1

Practice
Find the unknown value in each equation. The same shapes represent the same value.

7 $\square + \square = 7 + 5$

What is $\square$?

_____

8 $\bigcirc + \bigcirc = 14 - 6$

What is $\bigcirc$?

_____

9 $7 - \triangle = 3 + 4$

What is $\triangle$?

_____

10 $\diamond - 1 = 2 + 6$

What is $\diamond$?

_____

11 $\triangle + \triangle = 2(4 - 1)$

What is $\triangle$?

_____

12 $8 - \bigcirc = 6 - 3$

What is $\bigcirc$?

_____

13 $\square + \square + \square = 7 - 1$

What is $\square$?

_____

14 $\bigcirc + \bigcirc + \bigcirc = 3(4 + 2)$

What is $\bigcirc$?

_____

Reflect
What values of $\bigcirc$ and $\triangle$ make a true number sentence below? Is there more than one set of numbers that complete the number sentence? Explain.

$$\bigcirc - \triangle = 2(4 - 1)$$

Variables • Lesson 1

Week 1 Variables

Lesson 2

Key Idea
Variables can represent money amounts.

Try This

Ms. O'Brien works at the school bookstore. She sold the following items to students during lunch. Find the unknown costs of the items.

1. eraser + pencil = 40¢

   The eraser costs 25¢. What is the cost of the pencil?

   _____

2. marker + marker + folder = 95¢

   Each marker costs 30¢. What is the cost of the folder?

   _____

3. paper clip + paper clip + paper clip + paper clip + paper clip + paper clip + pen = 92¢

   Each paper clip costs 2¢. What is the cost of the pen?

   _____

Practice
Find each unknown cost of the items purchased from a hardware store.

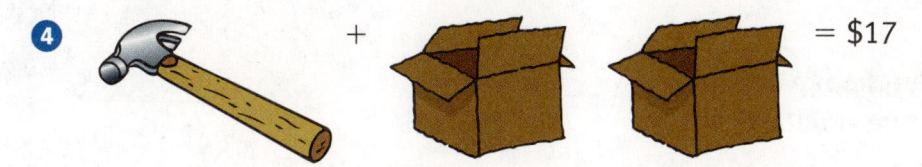

4 hammer + 2 boxes of nails = $17

Each box of nails costs $2.50. What is the cost of the hammer?

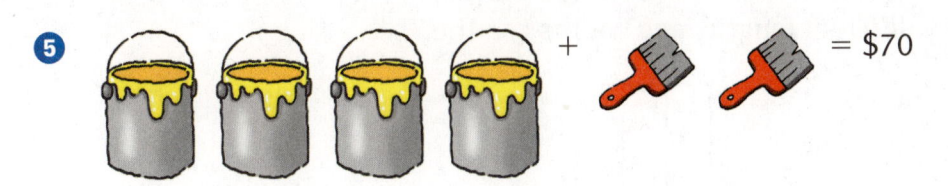

_____

5 4 cans of paint + 2 paintbrushes = $70

Each can of paint costs $15. What is the cost of each paintbrush?

_____

6 2 drills + 2 tape measures = $80

Each drill costs $32. What is the cost of each tape measure?

_____

Reflect
Suppose Carl spent $1.05 at the bookstore for two folders and one marker. If the cost of a folder is the same as the cost of a marker, what is the cost of each item? Explain.

_____

_____

Variables • Lesson 2

Week 1 Variables

Lesson 3

Key Idea
Variables stand for values that can change or vary. Some equations have more than one variable.

Try This

The area of a rectangle is given by the equation $A = l \times w$. To find the area, you multiply the length by the width. Suppose a rectangle has an area of 24 square units. How many different lengths and widths can the rectangle have?

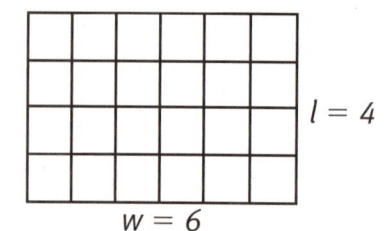

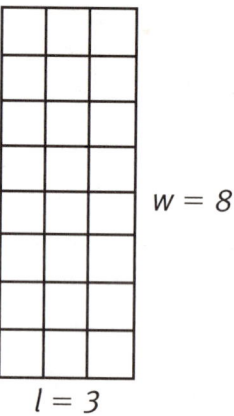

$l = 4$, $w = 6$

$w = 8$, $l = 3$

Complete the table for rectangles with an area of 24 square units. The two examples shown above have been completed in the table.

	$A = l \times w$	
	length, l	width, w
❶	1	
❷	2	
	3	8
❸		6
	6	4
❹		3
❺	12	
❻	24	

6 Number Patterns and Relationships • Week 1

Practice
The shapes represent two different variables. Complete the table for each equation.

$\bigcirc + \triangle = 15$

$\bigcirc$	$\triangle$
3	
5	
	9
	7
10	
	4

(7) (8) (9) (10) (11) (12)

$\square + \square + \triangle = 20$

$\square$	$\triangle$
2	
3	
	10
	8
7	
	2

(13) (14) (15) (16) (17) (18)

Reflect
What is another set of values for the square and triangle that would make a true number sentence?

Variables • Lesson 3   7

Week 1 — Variables

Lesson 4

Key Idea
In this lesson, you will continue to explore equations by using area models.

Try This
The model on the left shows a rectangle with an area of 63 square units. Use the model to answer each question.

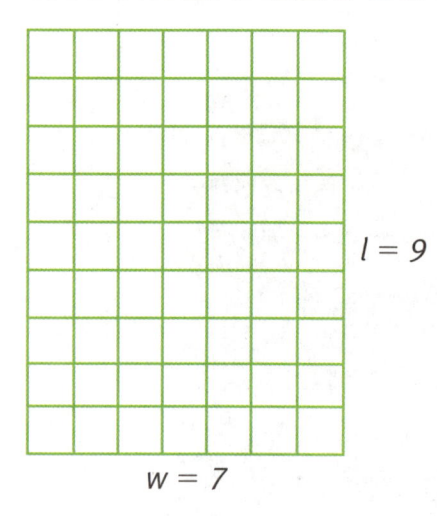

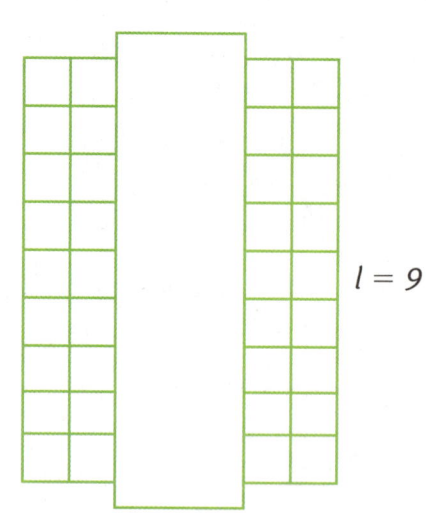

l = 9
w = 7

l = 9

① How many columns of squares are not hidden in the second figure?

_____

② Let △ represent the number of columns that are hidden in the second figure. Write an expression for the length.

_____

③ Write an equation for the area of the second figure with the variable △.

_____

8 Number Patterns and Relationships • Week 1

Practice

Write an equation for each area model where the length and width are known. Then write an equation using a variable to represent the hidden columns.

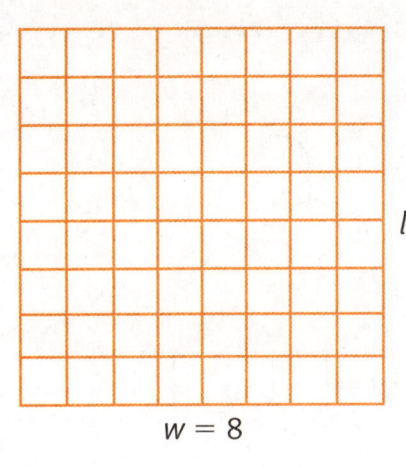

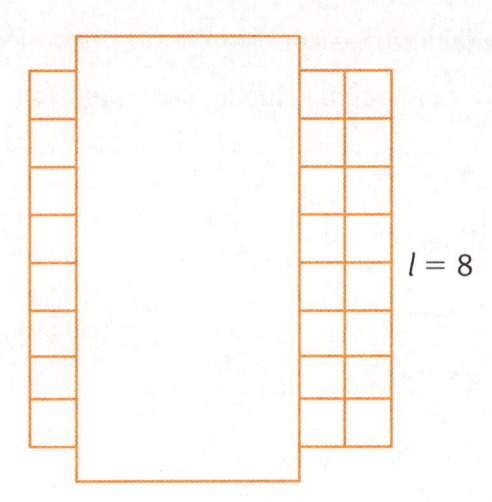

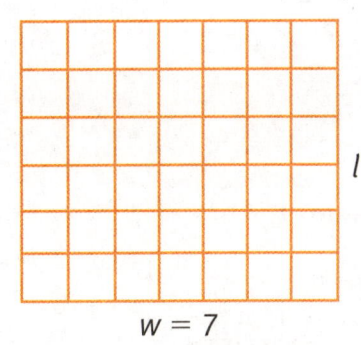

4) $l = 8$, $w = 8$

$l = 8$

5) $l = 6$, $w = 7$

$l = 6$

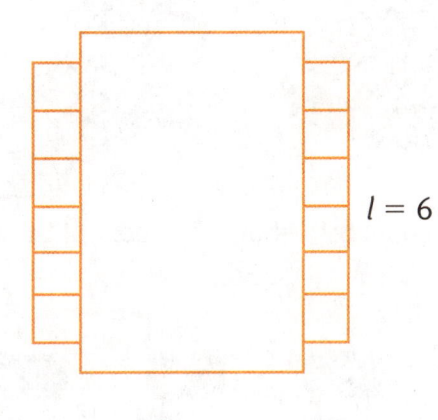

Reflect

How could you write an area equation from a model if a certain number of rows is unknown? Give an example of such an equation.

Variables • Lesson 4

Week 1 Variables

Lesson 5 Review

This week you used variables in equations. You chose values that could replace a variable and make a true number sentence.

Lesson 1 Find the unknown value in each equation.

1. $10 - 5 = \square + 2$
What is $\square$? _____

2. $11 + 1 = \bigcirc + 2$
What is $\bigcirc$? _____

3. $\triangle - 7 = 4 + 1$
What is $\triangle$? _____

4. $7 + 9 = \square + \square$
What is $\square$? _____

Lesson 2

5. 🍳 + 🧤 = $15

The pan costs $9. What is the cost of the two oven mitts?
_____

6. 🪑🪑🪑🪑 + 🪑 = $250

Each chair costs $25. What is the cost of the table?
_____

Reflect

To go along with the table and chairs, Susan purchased a fifth chair and a vase of flowers for the table. Her total purchase cost was $295. Write an equation to represent her purchase using *v* for the price of the vase. How much did the vase of flowers cost?

Lesson 3 Complete the table for each equation below.

$$\bigcirc + \triangle = 8 - 2$$

	$\bigcirc$	$\triangle$
7	1	
8	2	
9		3
10		0

Lesson 4 Write an equation for each area model where the length and width are known. Then write an equation using a variable to represent the hidden rows.

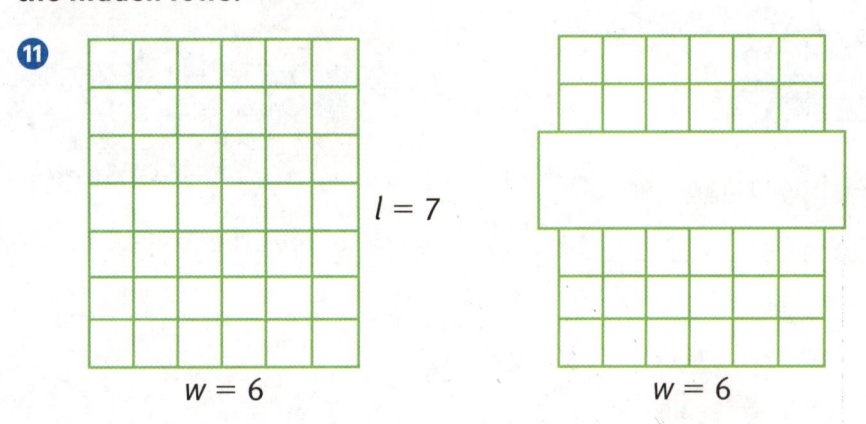

11

$l = 7$, $w = 6$

$w = 6$

Reflect

In Problem 11 how did you decide on which expression to use for the width of the rectangle that has hidden rows? How many rows are hidden? Show your work.

_____
_____
_____

Variables • Lesson 5 Review **11**

Week 2 Equality

Lesson 1

Key Idea
A seesaw, or teeter-totter, is a real world example of a balance scale. When a seesaw or balance scale is perfectly balanced, the items on each side have the same weight.

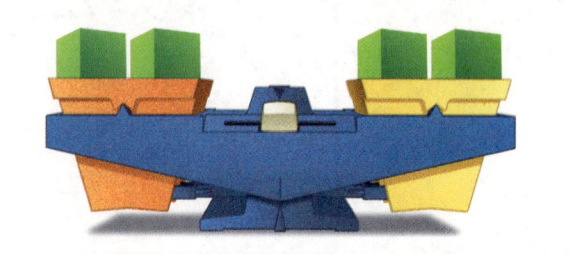

Try This
Use each balance scale to find two equal weights.

1

What does the weight of 1 apple equal?
_____

2

What does the weight of 1 flowerpot equal?
_____

3

What does the weight of 2 blocks equal?
_____

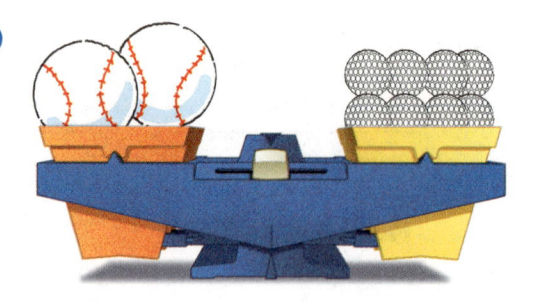

4

What does the weight of 1 baseball equal?
_____

12 Number Patterns and Relationships • Week 2

Practice
Tell how much each item might weigh for the seesaw to be balanced.

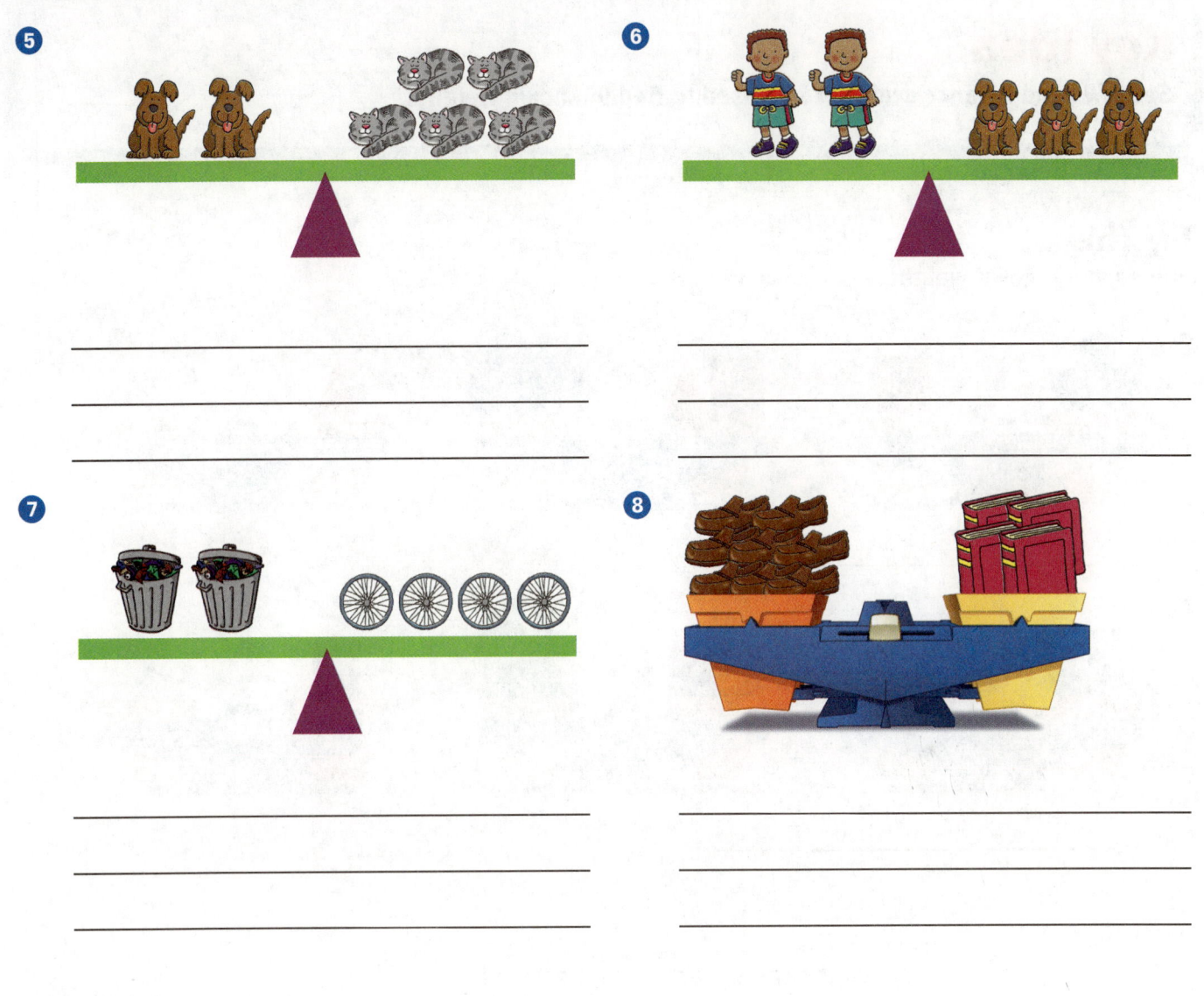

5.

6.

7.

8.

Reflect
Suppose Maria is holding her dog on one side of a seesaw. On the other side is her older brother Kyle. How much might Maria, the dog, and Kyle weigh for the seesaw to be balanced? Explain.

Equality • Lesson 1

Week 2 Equality

Lesson 2

> **Key Idea**
> Seesaws and balance scales can be used to find unknown weights.

Try This
Find each unknown weight.

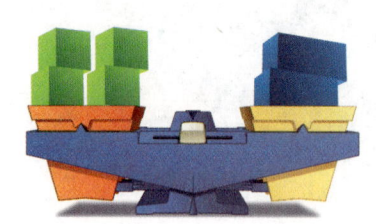

1

2

Practice
Which object(s) can you put on the third seesaw to balance it?

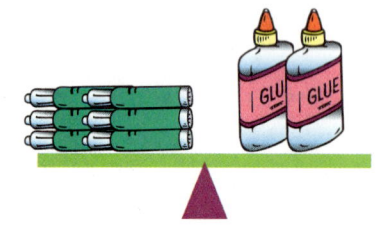

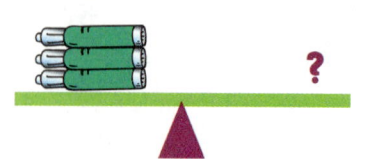

3

14 Number Patterns and Relationships • Week 2

④

⑤

_____

Reflect
In Exercise 5, explain how you determined your answer.

_____
_____
_____
_____

Equality • Lesson 2   15

Week 2 — Equality

Lesson 3

Key Idea
- A balance scale can be used to model equations and inequalities.
- If a scale is balanced, the two sides are equal, and it represents an equation.
- If a scale is not balanced, the left side is greater than (>) or less than (<) the right side. In this case, the scale represents an inequality.

$4 + 2 = 1 + 5$ $6 - 1 > 2 + 2$ $3 + 4 < 12 - 2$

Try This
Tell whether each scale is balanced. Write =, >, or < to make a true statement about each scale.

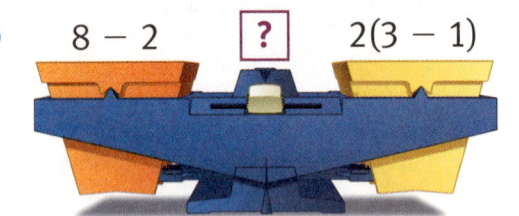

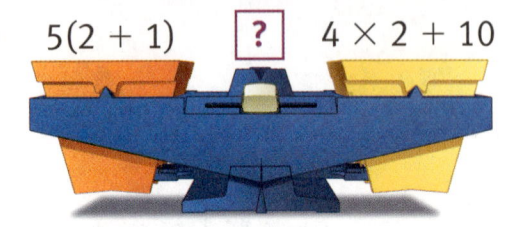

1 3×3 ? $4 + 5$ _____

2 $8 - 2$? $2(3 - 1)$ _____

3 $5(2 + 1)$? $4 \times 2 + 10$ _____

16 Number Patterns and Relationships • Week 2

Practice
Replace the question mark with the correct symbol to make a true statement.

4. 2 + 1 × 7   ?   3(4 − 2)   _____

5. 2(10 − 5)   ?   6 + 4   _____

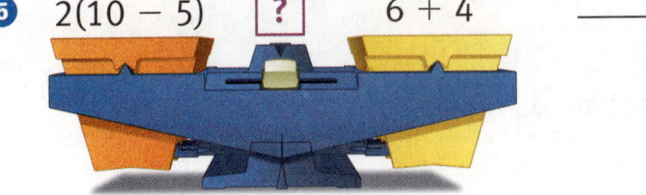

6. 4 × 3 − 6   ?   5(11 − 10)   _____

7. 4 + 2 × 3 − 1   ?   5(3 − 2) + 6   _____

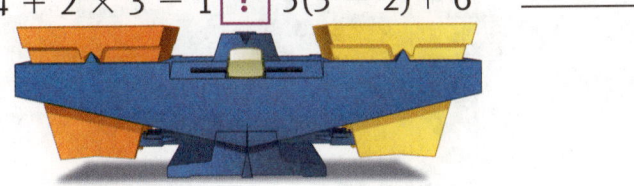

8. 4(8 − 6) + 3   ?   5 − (1 × 2) + 8   _____

Reflect
Write two expressions that would balance a scale when you put one on each side.

_____

Equality • Lesson 3

Week 2 — Equality

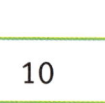

Lesson 4

Key Idea
- Balance scales can be used to model and solve equations.
- Each side of a balanced scale must have the same weight. You can use this information to help you set up an equation and solve for unknown values.

Try This
Fill in the table with values that will make the scale balanced.

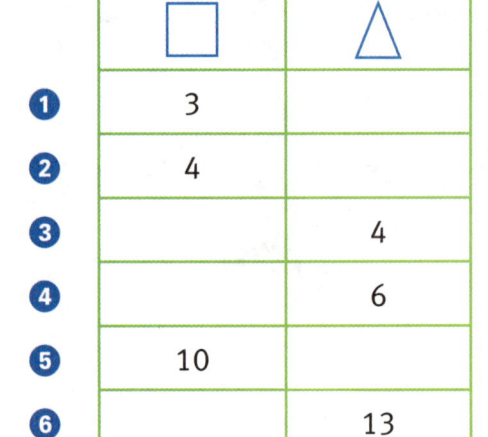

□ + 2 = △ + 4

	□	△
1	3	
2	4	
3		4
4		6
5	10	
6		13

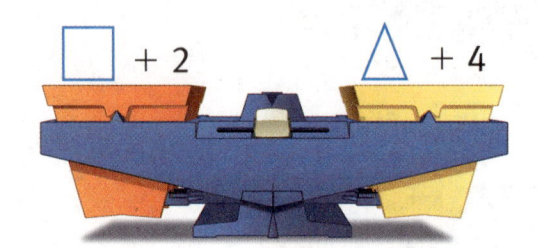

Practice
Write a number in each shape to balance the scale.

7. △ + 5 = ○ − 1

18 Number Patterns and Relationships • Week 2

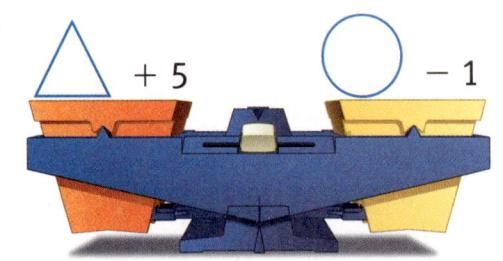

8 ☐ − 3 × 2   4 + ◯

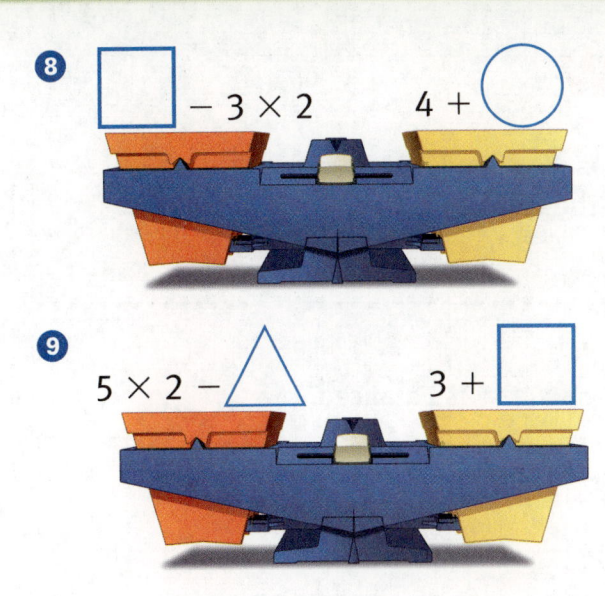

9 5 × 2 − △   3 + ☐

Fill in the table with values that will set the scale so the right side is greater than the left side.

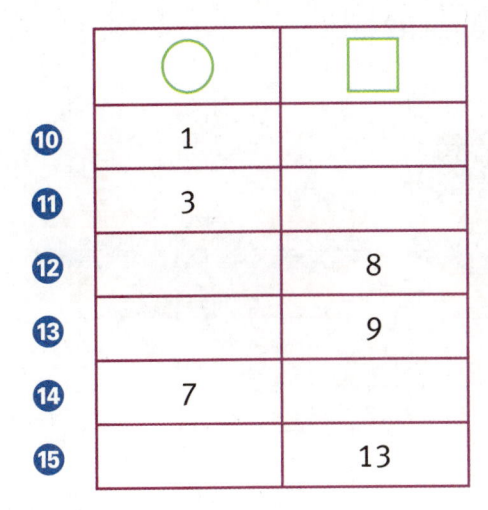

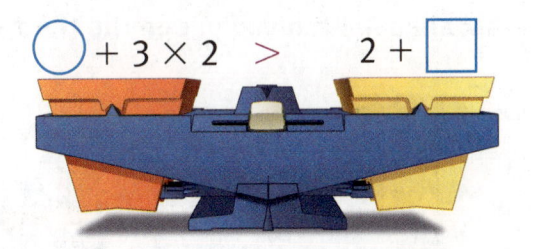

◯ + 3 × 2 > 2 + ☐

	◯	☐
10	1	
11	3	
12		8
13		9
14	7	
15		13

Reflect
Refer to the balance scale used in Problems 10–15. What numbers could you replace each variable with so the scale would represent a "<" inequality?

Equality • Lesson 4   **19**

Week 2 **Equality**

Lesson 5 Review

This week you explored balance in equations. You used balanced scales and seesaws to determine the value of a known object or expression.

Lesson 1 **Practice**
Tell how much each item might weigh so the seesaw is balanced.

❶

Lesson 2 What shape(s) can you put on the third scale to balance it?

❷

Reflect
Draw three balance scales like those in Problem 2. Use three different types of fruit. Write your answer, and show your work.

20 Number Patterns and Relationships • Week 2

Lesson 3 Replace the question mark with the correct symbol to make a true statement.

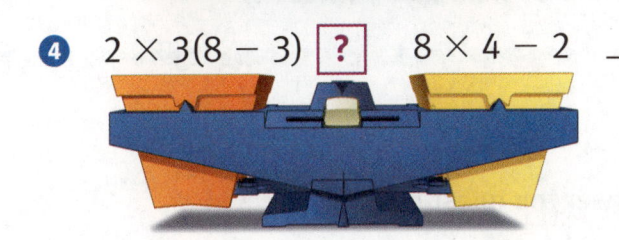

③ $3 + 2 \times 5$? $7(9 - 2)$ _____

④ $2 \times 3(8 - 3)$? $8 \times 4 - 2$ _____

Lesson 4 Fill in the table with values that will balance the scale.

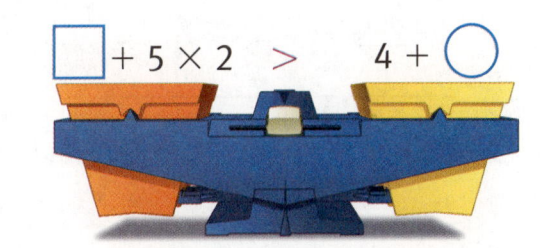

$\square + 5 \times 2 > 4 + \bigcirc$

$\square$	$\bigcirc$
1	
4	
	11
	12
8	
	16

⑤ 1
⑥ 4
⑦ 11
⑧ 12
⑨ 8
⑩ 16

Reflect

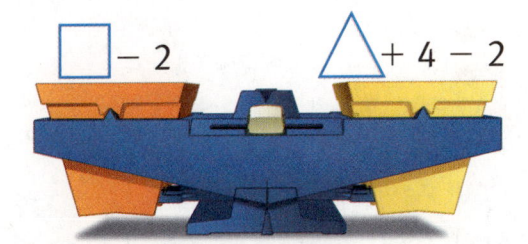

$\square - 2$ $\triangle + 4 - 2$

$\square$	$\triangle$

Create a table like the one in Problems 5–10 for the balance scale above.

Equality • Lesson 5 Review   21

Week 3

Functional Relationships

Lesson 1

Key Idea
- A function is a pattern in which each input value is paired with exactly one output value.
- These paired values can be organized into an input/output table.

Try This
Each carton of eggs contains 12 eggs. Answer the questions below.

1 How many eggs are there in 1 carton? In 2 cartons? In 3 cartons? Create an input/output table of this pattern. (Hint: The total number of cartons is the input. The total number of eggs is the output.

_____

2 Describe in words how you find the output if you know the input.

_____

3 What function rule could you use to find the number of eggs in *c* cartons?

_____

4 Suppose you know the total number of eggs (output). Describe in words how you could find the number of cartons.

_____

5 Complete the rows of the input/output table.

Input (number of cartons)	Output (number of eggs)
1	12
2	
3	
4	

22 Number Patterns and Relationships • Week 3

Practice

Lucy purchased a bus pass for $20. Each time she rides the bus, $0.50 is deducted from the balance on the pass.

6 How much will the balance on the pass be after Lucy rides the bus 1 time? After 2 times? After 3 times?

7 Use words to describe how you could find the output if you know the input.

8 Use words to describe how you would determine how many rides Lucy has taken (input) if you know the balance on the pass.

9 Complete the rows of the input/output table.

Input (number of bus rides)	Output (bus-pass balance)
0	$20.00
1	
2	
3	
4	

10 What function rule could you use to find the balance remaining after r rides?

Reflect

How many times can Lucy ride the bus before the balance reaches 0? Explain.

Functional Relationships • Lesson 1   23

Week 3

Functional Relationships

Lesson 2

Key Idea
- Patterns are represented using pictures, words, tables, rules, and graphs.
- You can create graphs of functions from input/output tables by plotting the ordered pairs.

Try This
Follow the steps to create a graph of the egg function from the previous lesson.

Input (number of cartons)	Output (number of eggs)
1	12
2	24
3	36
4	48
5	60
6	72

Step 1 Label the horizontal axis and the vertical axis.

Step 2 Plot a point for each ordered pair of numbers in the table.

Step 3 Give your graph a title.

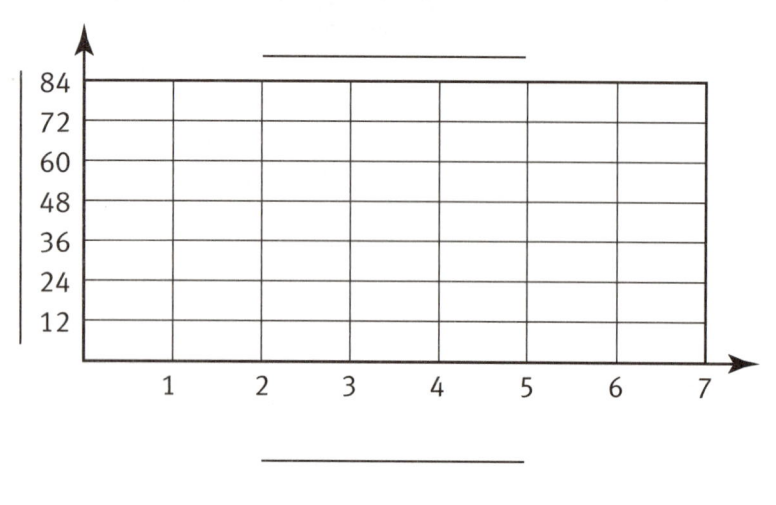

24 Number Patterns and Relationships • Week 3

Practice
Create a graph for the bus-pass function.

Input (number of bus rides)	Output (bus-pass balance)
0	$20.00
1	$19.50
2	$19.00
3	$18.50
4	$18.00
5	$17.50

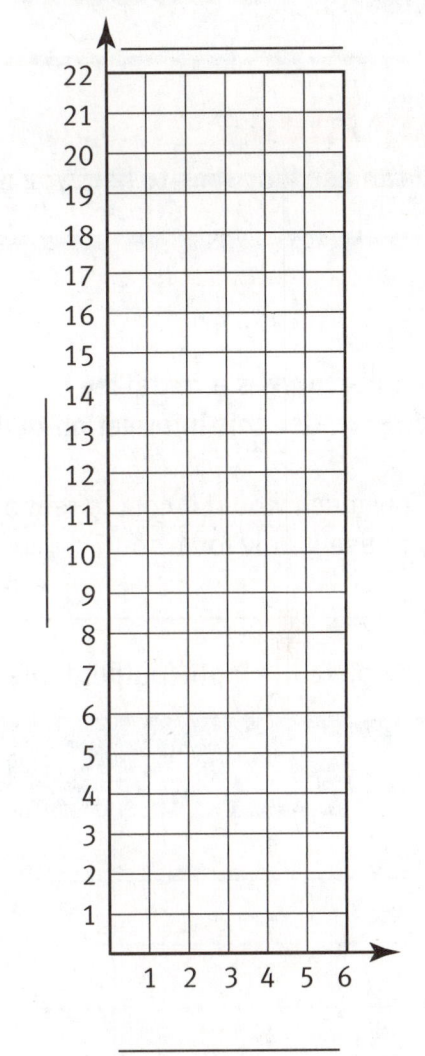

Reflect
Do you think it is possible to determine a function rule just by looking at the graph? Explain and give an example.

____

Functional Relationships • Lesson 2   25

Week 3

Functional Relationships

Lesson 3

Key Idea
You can use functions to help you make decisions.

Try This
E-Z Rentals charges a rental fee of $15 plus $5 per hour to rent a chain saw. Use this information to answer each question below.

1 How much would it cost to rent a chain saw if you use it only for 1 hour?

_____

2 How much would it cost to rent a chain saw if you use it for 2 hours?

_____

3 Complete the input/output table.

Input (number of hours)	Output (total cost)
1	
2	
3	
4	
5	
6	
7	
8	

4 Describe in words how you could determine the total cost (output) if you are given the number of hours the chain saw is rented (input).

_____

5 Write a function rule that can be used to find the total cost of renting the chain saw for h hours.

_____

26 Number Patterns and Relationships • Week 3

Practice
Use your answers from Try This to solve each problem.

6 Graph the data from your input/output table. Include points for renting the chain saw for up to 12 hours.

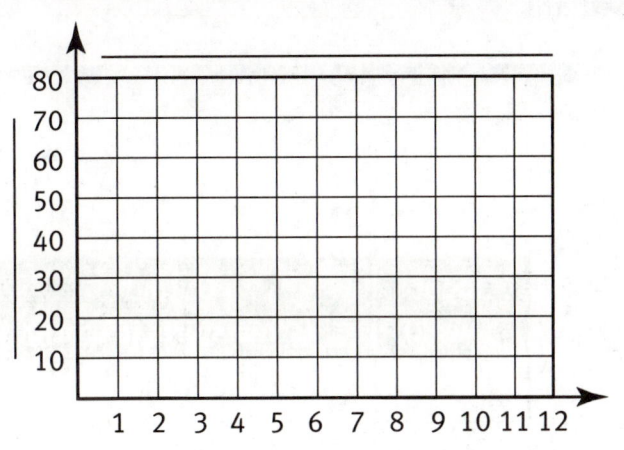

```
80
70
60
50
40
30
20
10
    1  2  3  4  5  6  7  8  9  10 11 12
```

7 Suppose the same chain saw can be purchased for $135. How many hours would you need to use the saw for it to be a better bargain to purchase instead of rent? Explain.

_____

_____

Reflect
If the dots in the graph in this lesson were connected, what would you see? Why do we not connect the dots?

_____

_____

_____

_____

Functional Relationships • Lesson 3

Week 3

Functional Relationships

Lesson 4

Key Ideas
You can use graphs to compare two related patterns.

Try This

Create a graph for each input/output table.

Input (week)	Output (sweatshirts sold this year)
1	15
2	30
3	45
4	60
5	75

Input (week)	Output (sweatshirts in stock)
1	100
2	85
3	70
4	55
5	40

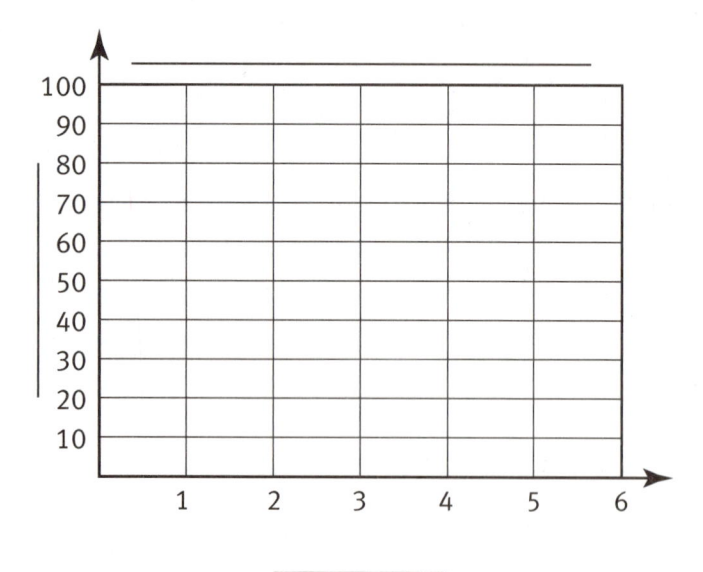

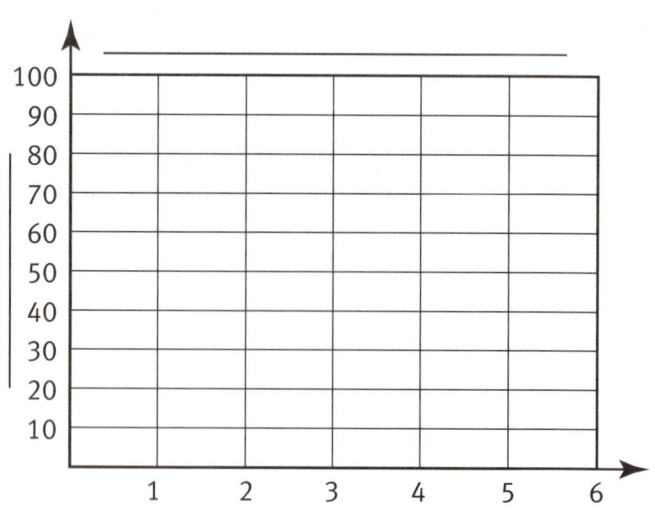

28 Number Patterns and Relationships • Week 3

Practice
Use your graphs from Try This to answer each question.

1 Describe the pattern shown in the first graph.
_____

2 Describe the pattern shown in the second graph.
_____

3 Which graph shows a growing pattern?
_____

4 Which of the graphs shows a shrinking pattern?
_____

5 What stays the same in the first graph?
_____

6 What stays the same in the second graph?
_____

7 How are the two graphs related?
_____
_____

Reflect
Suppose you are in charge of ordering sweatshirts for the store. How many sweatshirts would you want to have in stock for an 8-week period? Explain.
_____
_____
_____
_____
_____

Functional Relationships • Lesson 4

Week 3

Functional Relationships

Lesson 5 Review

This week you explored functions and function patterns. You studied input/output tables and graphed the data from input/output tables.

Lesson 1 Tickets to the state fair cost $15 each.

❶ Complete the input/output table.

Input (number of tickets)	Output (total cost)
1	
2	
3	
4	
5	

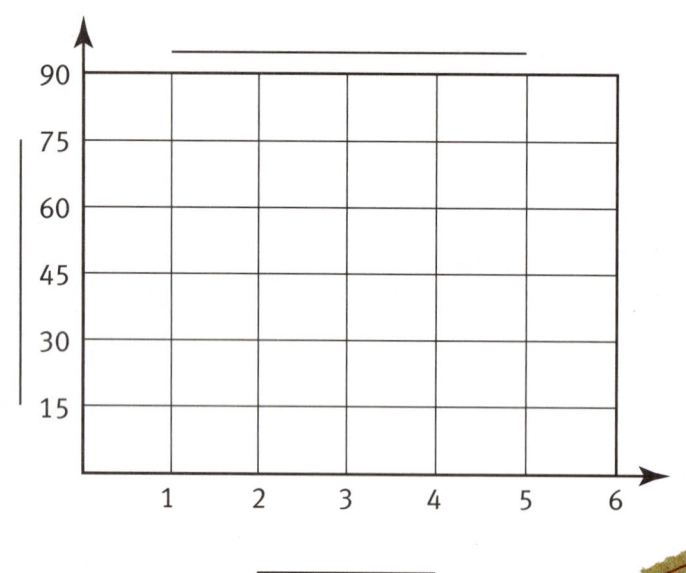

Lesson 2 ❷ Graph the data from the input/output table above.

Reflect
What function rule can be used to find the total cost of t tickets from the function above?

30 Number Patterns and Relationships • Week 3

Lesson 3

3. Graph the pattern shown in the input/output table.

Input (number of packages)	Output (hot dog buns)
1	8
2	16
3	24
4	32

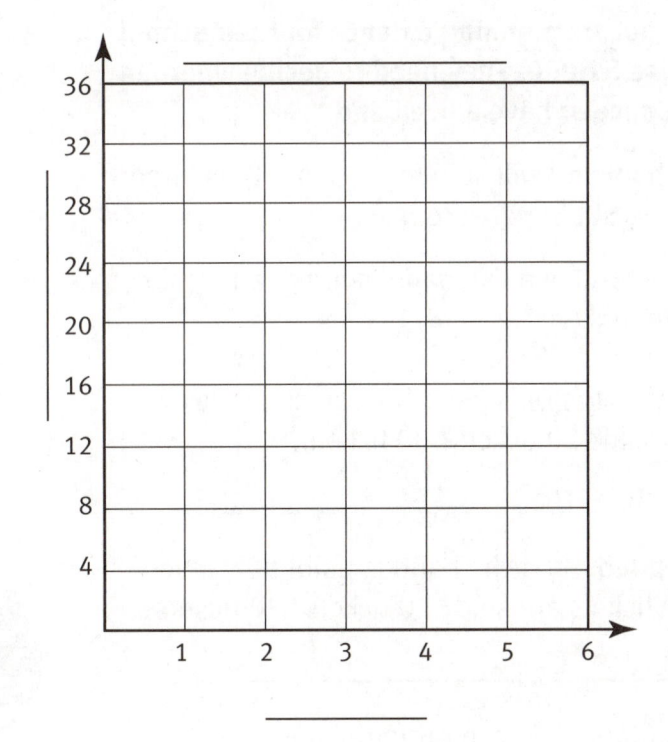

Lesson 4

4. How would you describe the pattern shown in the graph?

_____

5. What function rule can you use to find the number of hot dog buns in p packages?

_____

Reflect

Suppose Monica needs 150 hot dog buns for a reception. How many packages of buns should she buy?

_____

Functional Relationships • Lesson 5 Review

Week 4

More with Functional Relationships

Lesson 1

Key Idea
You can compare two functions and use them to help you make a decision.

Try This
Students at Richfield High School are planning a dance for their school fund-raiser. Their goal is to raise $1,000. They need to decide whether they want to hire a DJ for the dance or have a live band.

- The DJ is a student at the school and will spin for $75. If a DJ is hired, admission to the dance will be $7.50 per student.

- It will cost $250 to hire a live band. If a live band is hired, admission to the dance will be $10 per student.

1 Suppose the DJ is hired. After the DJ is paid, how much money will be raised if 10 tickets are sold? 20 tickets? 30 tickets?

_____

2 Suppose the live band is hired. After the band is paid, how much money will be raised if 10 tickets are sold? 20 tickets? 30 tickets?

_____

3 Complete the input/output table for each option.

| DJ Option ||
Input (tickets sold)	Output (money raised)
10	$0
20	
30	
40	
50	
60	

| Live Band Option ||
Input (tickets sold)	Output (money raised)
10	
20	
30	
40	
50	
60	

Practice
Use the information from Try This to answer each question.

4 Describe the pattern shown in the input/output table for the DJ option.
___
___

5 Describe the pattern shown in the input/output table for the live band option.
___
___

6 Suppose the DJ is hired. What function rule shows the amount of money raised if *t* tickets are sold?
___

7 Suppose the live band is hired. What function rule shows the amount of money raised if *t* tickets are sold?
___

8 Suppose the dance committee expects to sell 150 tickets for the dance. Which option should they choose to raise the most money for the school?
___

Reflect
How much money would be raised if 150 tickets were sold and the DJ option were used? How much money would be raised if 150 tickets were sold and the live band option were used?
___
___
___
___

More with Functional Relationships • Lesson 1

Week 4

More with Functional Relationships

Lesson 2

Key Idea
You can represent a geometric pattern as a function.

Try This
Grandma Rawls makes patchwork quilts. Each quilt has a row of yellow patches around the perimeter. In the center of the quilts are square red and white patches. The first four sizes of quilts are shown below.

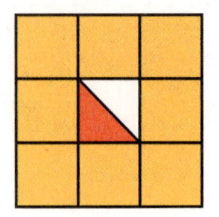

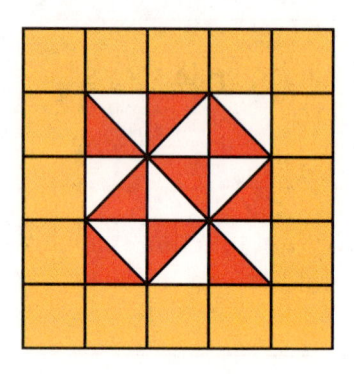

Size 1   Size 2   Size 3   Size 4

❶ How many square patches are used in each of the first three sizes?
_____

❷ How many yellow patches are used in the borders of each size?
_____

34   Number Patterns and Relationships • Week 4

Practice
Use the quilt designs to answer each question.

③ Complete the input/output table for the number of yellow squares used to create the border of each quilt.

Input (size)	Output (yellow border patches)
1	8
2	
3	
4	
5	
6	

④ What pattern do you notice in the table?

_____
_____

⑤ What function rule can be used to find the number of yellow border patches needed for a size n quilt?

_____

⑥ How many yellow border patches would Grandma Rawls need for a size-10 quilt?

_____

Reflect
What size quilt would have 36 yellow border patches? Show your work.

_____
_____
_____
_____

Week 4 — Lesson 3

More with Functional Relationships

> **Key Idea**
> You can represent a geometric pattern as a function.

Try This

Use the quilt patterns again to answer each question.

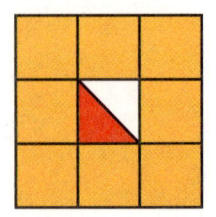

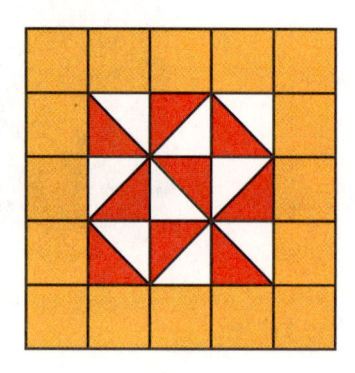

Size 1   Size 2   Size 3   Size 4

① How many patches are used in each of the first four sizes?

_____

② How many red/white square patches are used in each size?

_____

36 Number Patterns and Relationships • Week 4

Practice
Use the quilt designs to answer each question.

3 Complete the input/output table for the number of red/white squares used to create each quilt.

Input (size)	Output (red/white patches)
1	1
2	
3	
4	
5	
6	

4 What pattern do you notice in the table?

_____

_____

5 What function rule can be used to find the number of red/white patches needed for a size n quilt?

_____

6 How many red/white patches would Grandma Rawls need for a size-10 quilt?

_____

Reflect
What size quilt would have 144 red/white patches? Show your work.

_____

_____

_____

_____

More with Functional Relationships • Lesson 3

Week 4

More with Functional Relationships

Lesson 4

Key Idea
Relationships exist between equations and graphs.

Try This
Complete the input/output table for the equation below.

$\triangle = \bigcirc + 3$

Input $\bigcirc$	Output $\triangle$
1	
2	
3	
4	
5	
6	

1 Plot the values from the table on the coordinate grid.

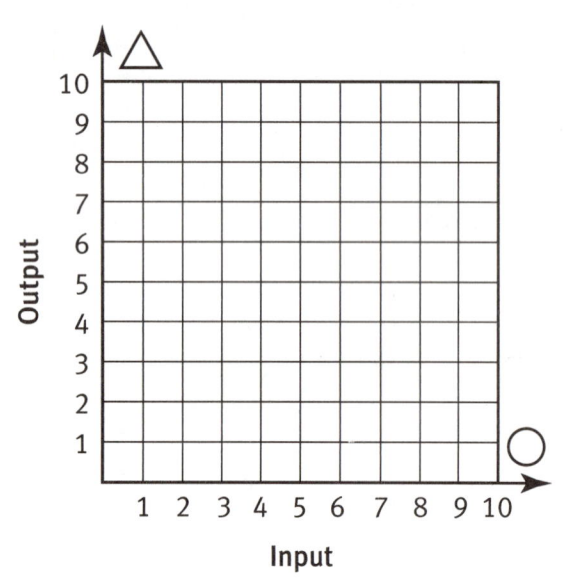

38 Number Patterns and Relationships • Week 4

Practice
Complete the input/output table for the equation below.

△ = ○ − 3	
Input ○	Output △
3	
4	
5	
6	
7	
8	

2 Plot the values from the table on the coordinate grid.

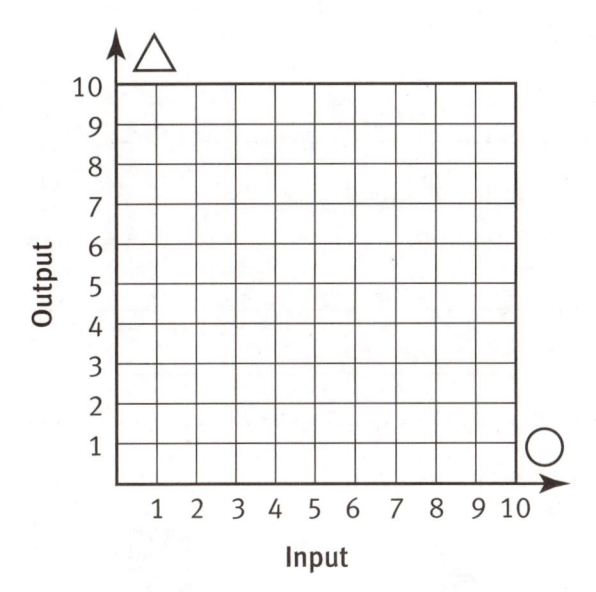

Reflect
Describe the similarities and differences between the two graphs in this lesson.

_____
_____
_____

More with Functional Relationships • Lesson 4

Week 4 — More with Functional Relationships

Lesson 5 Review

This week you learned more about functional relationships. You discovered that geometric patterns can be represented as functions. You learned that relationships exist between equations and graphs.

Lesson 1 Use the table below to answer each question.

Car Rental	
Company A	**Company B**
• Up-front fee: $25	• Up-front fee: $0
• Rental fee: $15/day	• Rental fee: $20/day

❶ Write a function rule for the total cost of renting a car from Company A for d days.

_____

❷ Write a function rule for the total cost of renting a car from Company B for d days.

_____

❸ Suppose Miss Wilson needs to rent a car for 7 days. Which company will be less expensive? Explain.

_____
_____

Lesson 2 Use the pattern of blocks below to answer each question.

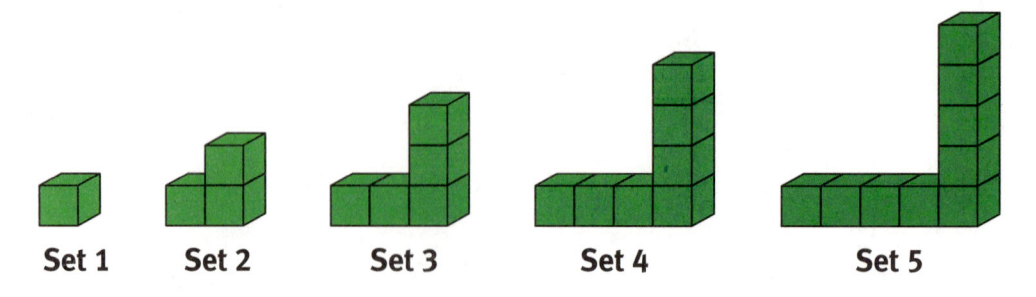

Set 1   Set 2   Set 3   Set 4   Set 5

❹ How many blocks were used to create each of the five figures?

_____

40 Number Patterns and Relationships • Week 4

Lesson 3 Use the pattern of blocks below to answer each question.

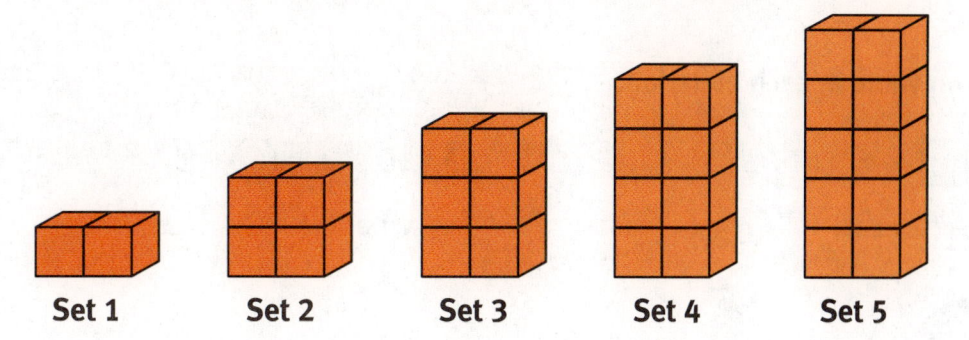

Set 1   Set 2   Set 3   Set 4   Set 5

5 How many squares were used to create each of the five figures?
_____

6 Write a function rule that tells how many blocks are in the *n*th set of the pattern.
_____

Lesson 4 **7** Complete the input/output table for the equation below. Then plot the values on the coordinate grid.

◯ + 2 = □

Input ◯	Output □
1	
2	
3	
4	
5	
6	

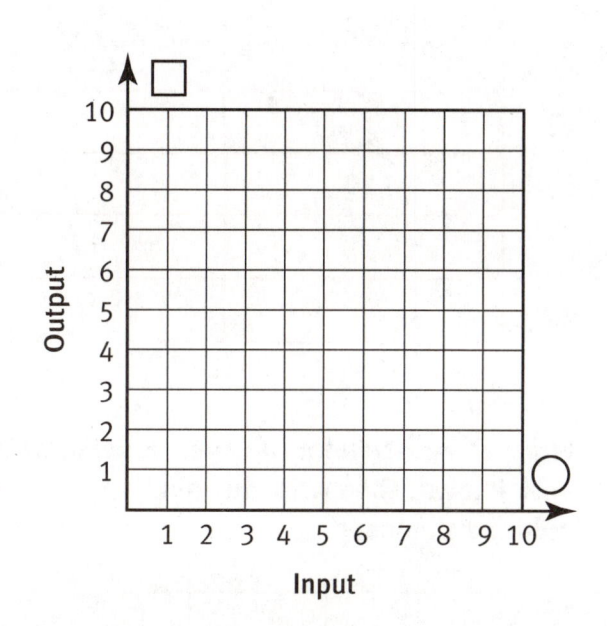

Reflect
Does the graph above show all the answers to this equation? Explain.
_____
_____

More with Functional Relationships • Lesson 5 Review   **41**

Week 1

Variables

Practice

Find the unknown value in each equation.

1. $19 - 3 = \square + 4$

What is $\square$? _____

2. $16 + 2 = \bigcirc + 5$

What is $\bigcirc$? _____

3. 🌂 + ⌚ = $34

The umbrella costs $15. What is the cost of the watch? _____

Complete the table for the equation below.

$\bigcirc - \triangle = 9 - 4$	
$\bigcirc$	$\triangle$
4. 6	
5. 20	
6.	19
7.	36

Write an equation for the area model in which the length and width are both known. Then write an equation using a variable to represent the hidden columns.

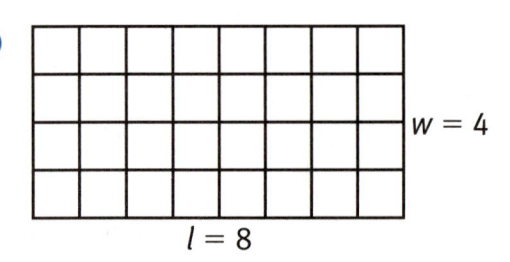

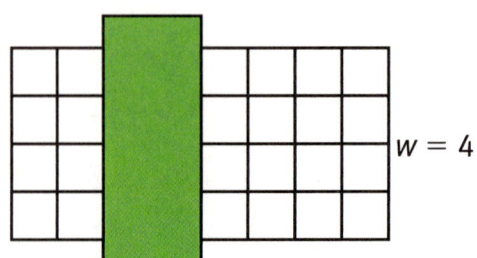

8. $l = 8$, $w = 4$ $w = 4$

_____

42 Number Patterns and Relationships • Week 1 Practice

Week 2 Equality

Practice

Tell how much each item might weigh so that the seesaw is balanced.

1

What shape(s) can you put on the third scale to balance it?

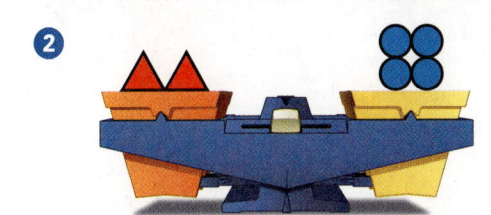

2

_____

Replace the question mark with the correct symbol to make a true statement.

3 $3 + 9 \times 3$ **?** $9(6 - 1)$ _____

Fill in the table with values that will balance the scale.

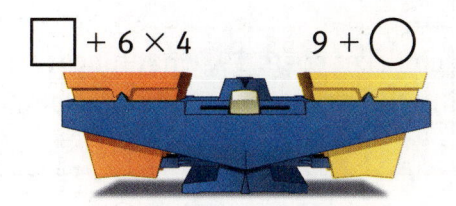

	□	○
4	4	
5	8	
6		15
7		20
8	9	

$\square + 6 \times 4 \quad = \quad 9 + \bigcirc$

Number Patterns and Relationships • Week 2 Practice   43

Week 3 — Functional Relationships

Practice

1 Movie tickets to the cinema complex cost $8 each. Complete the input/output table.

Input (number of tickets)	Output (total cost)
1	
2	
3	
4	
5	

2 Graph the data from the input/output table.

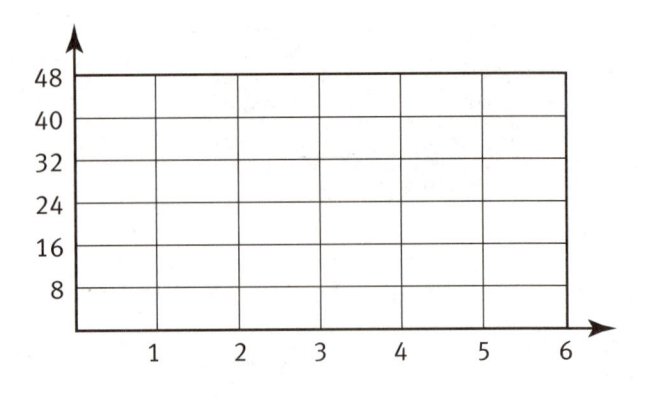

Use the graph to answer the questions below.

3 How would you describe the pattern shown in the graph?

_____

4 What function rule can you use to find the cost of *t* tickets?

_____

5 How much was spent if you have 12 tickets?

_____

6 Is the graph showing a shrinking or growing pattern?

_____

7 If each ticket costs $9, what function rule can you use to find the amount spent on *t* tickets?

_____

Week 4

More with Functional Relationships

Practice

Use the table below to answer each question.

Hotel Conference Room Booking Cost	
Hotel A	**Hotel B**
• Up-front fee: $120 • Usage fee: $25/hour	• Up-front fee: $0 • Usage fee: $45/hour

1 Write a function rule for the total cost of booking a conference room at Hotel A for *h* hours, then write a similar function for Hotel B.

_____

2 Suppose Mr. Fraley needs to book a conference room for 8 hours. Which hotel will be less expensive? Explain.

_____

_____

3 Write a function rule that tells how many octagons are in the *n*th term of the pattern.

Set 1   Set 2   Set 3   Set 4

_____

4 Complete the input/output table for the equation below. Then plot the values on the coordinate grid.

○ + 2 = □

Input	Output
1	3
2	4
3	5
4	6
5	7
6	8

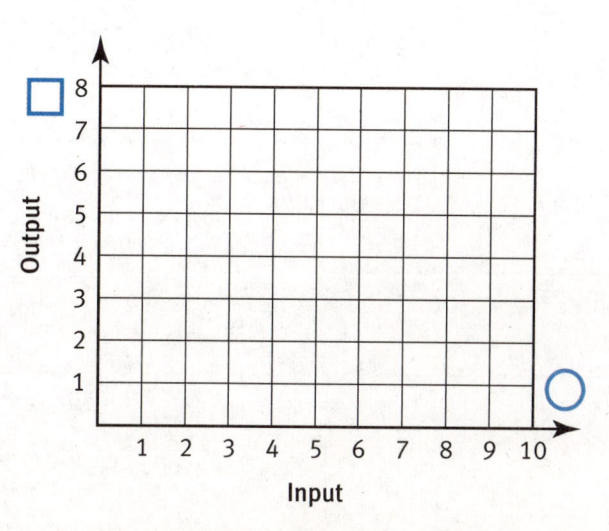

Number Patterns and Relationships • Week 4 Practice

SRA Number Worlds

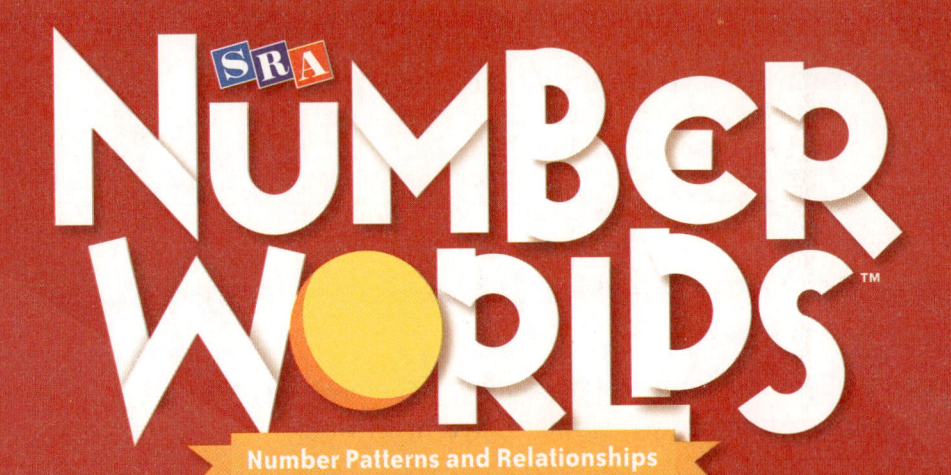

Number Patterns and Relationships

Unit 2 Workbook

SRAonline.com

The McGraw·Hill Companies

Level G R53245.01